2030
The Year Civilisation Will Die

How We Can Fix the Triad of

Global Warming,

Population Collapse
and
Antibiotic Failure

Only farmers
& foresters
can do this!

Bill Butterworth

Jan '20

2030 - The Year Civilisation Will Die
First published in 2019 by
Acorn Books
www.acornbooks.co.uk

Acorn Books is an imprint of
Andrews UK Limited
www.andrewsuk.com

Contents

Managing the Triad iv

Introduction v

Section 1
Chapter 1: The Report and You. .2
Chapter 2: Other Global Research 5
Chapter 3: A Discussion About Pollution 9

Section 2
Chapter 4: Population, People and Problems 20
Chapter 5: Conclusions So Far 37

Section 3
Chapter 6: Why Do Anything? 42
Chapter 7: Electric Cars and Nuclear Power 43
Chapter 8: Transition Fuels 45
Chapter 9: Sustainable Energy Production. 55
Chapter 10: Sustainable Oceans 71

Section 4
Chapter 11: Soils and Greenhouse Gases 74
Chapter 12: How The Closed Loop Really Works. 77
Chapter 13: Soil Carbon, Bio-Active Soils & BACS 83
Chapter 14: Using Wastes to Feed Photosynthesis 85
Chapter 15: Which Wastes 108
Chapter 16: Global Scale – Quantities & Qualities 128
Chapter 17: Global Scale – Organisation & Reverse Franchising . . 133
Chapter 18: Sub-Surface Reservoirs 136

Section 5
Chapter 19: Sustainability Notes 140
Chapter 20: Farming and Forestry for 2030 151
Chapter 21: Top Soil Reservoirs. 173

Section 6
Chapter 22: The Part of Government 184

Section 7
Chapter 23: Discussion & Conclusions. 190

Final Thought. 195

Also from Bill Butterworth. 196

Managing the Triad

This book is not about global warming and whether it is taking place or not, how fast it is and what effects it will have. Nor is it about population growth. Nor is it about running out of effective vaccines and antibiotics.

Those issues, in isolation, are covered in great depth by many other sources. This book brings the three together, considering the possibility of the triad becoming critical *at the same time*. Experts disagree on exact dates, but the most likely period this will happen to us is some time between 2030 and 2040 if we are lucky, a little earlier if not!

If and when this three-fold tipping point occurs, there will only be a short period before a major collapse in global population, a humanitarian disaster on a scale of unprecedented magnitude.

This book is not designed to scare or frighten the reader, but to help them consider the true risk of this upcoming catastrophe. Once understood, it then details a powerful solution available to us with current technology: the global Bio-Active Carbon Sink.

Introduction

This book is structured as follows:

Section 1 introduces the first and very pressing part of the triad; the IPCC report into global warming and the pollution of our environment. Carbon dioxide is by far the biggest single contributor to global warming, and this section describes why we need to take it out of the atmosphere.

Section 2 investigates the second and third elements of the triad: population growth and the imminent failure of our current spectrum of antibiotics.

Section 3 is concerned with our energy requirements, considering the options available to us at the current time.

Section 4 introduces the reader to the concept of the Bio-Active Carbon Sink. It looks into the synergy of getting back the oxygen from the carbon dioxide in our atmosphere. Other considerations within the section include safer organic food and the general pollution within our environment.

Section 5 develops the practicalities of recycling to land, and considers the ongoing maintenance of the synergy introduced in the previous section.

Section 6 is concerned with leadership; how the necessary objectives can be realistically achieved through regulations introduced by governments around the world.

Section 7 asks what we can truly believe, what we can conclude, and what the likely time-scales will be.

To round off this brief introduction, I present you firstly overleaf with a figure that demonstrates how such a catastrophic population collapse is a real possibility based on the IPCC report. Then, before we begin in detail, I have included a quote from the late Professor Stephen Hawking that I found to be of some relevance:

THE CATASTROPHIC POPULATION CURVE

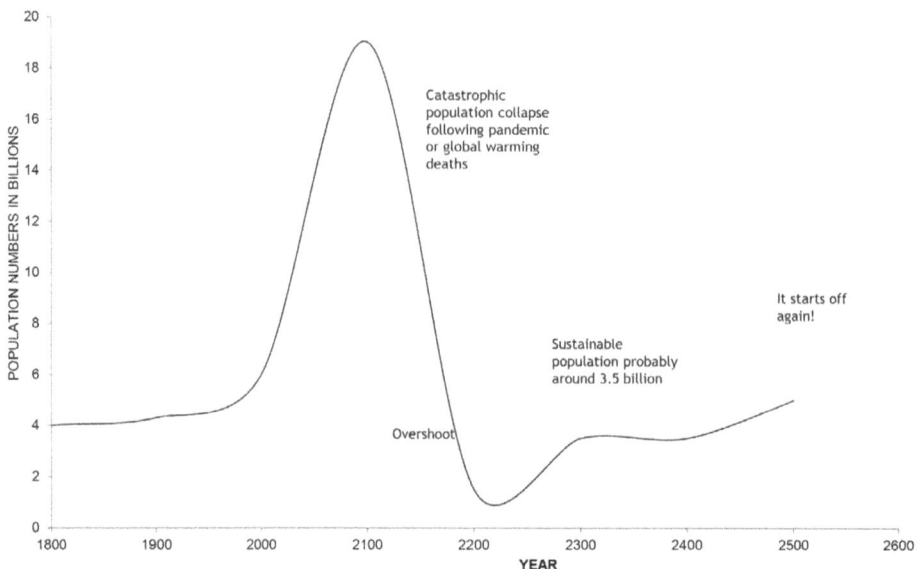

There are big questions to be answered. How will we feed an ever-increasing population? Provide clean water, generate renewable energy, prevent and cure disease and slow down global climate change?
—Stephen Hawking

2030
The Year Civilisation Will Die

Section 1

The IPCC report of October 2018

We have 13 years to save the planet
—The IPCC

Chapter 1: The Report and You

Who Are These People?

IPCC is the Intergovernmental Panel on Climate change.[1] It is just about the most informed, most independent, least emotional, body of scientist ever to express an opinion about climate. These are the top guys in the world and as good as we can get, for the present at least.

Created in 1988 by the World Meteorological Organization (WMO) and the United Nations Environment Programme (UNEP), the objective of the IPCC is to "provide governments at all levels with scientific information that they can use to develop climate policies". Clever people with no particular axe to grind. Just giving the facts as they are. Hundreds of the world's leading experts volunteer their time to survey the available research and write reports, hundreds more check the drafts and offer advice. Note the word *'volunteer'*. While nothing in this world is absolute, these reports are about as independent, transparent and truthful as it is possible to get. While it has to be remembered that the latest report looking at the potential rise in global temperatures is about predicting the future (and there is nothing certain about that), it is very unlikely that these people are on the wrong track.

What is the *Tipping Point?*

During the summer of 2018, the IPCC went through more than one draft of a report which was finally issued in the autumn. When it was finally understood, the news began to grab headlines. The report's simple title was; "*An IPCC special report on the impacts of global warming of 1.5°C above pre-industrial levels and related global greenhouse gas emission pathways, in the context of strengthening the global response to the threat of climate change, sustainable development, and efforts to eradicate poverty.*" By any standards, any logic, relating climate change to sustainable development is difficult enough but to add in '*to eradicate poverty*' is a tall order. At least to *start* delivering... that is what this book is attempting to contribute to.

Why *tipping point*? Well, many natural phenomena are very slow to start developing and they gradually gain speed and inertia. Eventually, they get to a point where whatever is done to stop it, it has enough inertia to keep going. The gradual curve upwards on the graph slowly but inexorably becomes exponential and gets out of control. What is argued here is that global temperatures have risen by 1°C above pre-industrial levels and are expected to rise more rapidly from now on, and that this rate of rise may become unstoppable if it gets to another 1.5°C by 2030. These situations are complex but, to give an example of how this tipping point might work, look at the oceans and their carbon dioxide content. As we burn fossilised fuel, some of the carbon dioxide produced is absorbed by sea water, about a quarter of it all. When that happens, the seawater becomes more acidic and the corals and many other sea creatures fail to

---

1.   You can look them up at https://www.ipcc.ch/about/

adapt and die (We know that is already happening). However, as global warming builds, the sea water begins to release some of its carbon dioxide and that goes back into the atmosphere and creates more global warming. That, then, gets us closer to the tipping point where the exponential rise takes over and, at some point, there may be nothing we can do about it. That is the tipping point.

As another example, as the global air temperature rises, so does the ground close to the poles, i.e. what is known as the areas of *permafrost*. When these melt, they will release billions of tonnes of carbon dioxide into the atmosphere and global warming will suddenly speed up – another tipping point.

These self-reinforcing feedbacks which, if allowed to continue, will accelerate warming and risk cascading climate tipping points and runaway warming. The IPCC report came to the conclusion that unless we limit further global temperature rise to no more than 1.5°C, then the danger of a tipping point kicking in was very likely. There are many possible tipping point mechanisms. For example, when ice melts at the poles it is known to be of particular danger; the earth's ice caps act as reflectors, sending some of the sun's rays back into space and cooling the planet. When sea ice melts, it reveals dark water underneath, which absorbs more heat and in turn triggers greater warming of the water, releasing more carbon dioxide, which causes more atmospheric warming, which melts more ice, in a constant feedback loop.

What It Almost Certainly Means for Our Kids

Even if we do manage to limit global temperature rises to another 1.5°C, there are still very serious consequences of environmental changes. However, there are many scientists who take the view[2] that the IPCC report underestimated the risks of triggering more than one tipping point and the onset of run-away global warming poses potential catastrophic consequences to not only human civilisations but also wild life in general.

There is a more insidious threat, however. We are polluting the environment we live in with chemicals which will complicate our health and kill many of us. As an example beyond such things as industrial and traffic pollutants which have been given plenty of publicity elsewhere, sometimes for many years, recent research shows evidence that pharmaceutical drug residues which have passed through the human body are significantly worse than previously, commonly understood. Australian research[3] identified sixty-nine named drugs which had got through the sewage collection and treatment systems into river bed muds and back up the food chain. In the countries with higher population density and greater average age, than Australia, the risks are likely to be higher and more complex. For this problem, too, there is a potential solution in the soil carbon sink.

2. Harvey, F: *Tipping Points Could Exacerbate Climate Crisis, Scientists Fear, The Guardian*, October 9th 2018 https://www.theguardian.com/environment/2018/oct/09/tipping-points-could-exacerbate-climate-crisis-scientists-fear

3. Klein, A: *More Than Sixty Prescription Drugs Are Getting Into River Foodchains, New Scientist* Oct 6th 2018 https://www.newscientist.com/article/2184420-more-than-60-prescription-drugs-are-getting-into-river-foodchains/

What It Means for Us, Right Now

It is up to each one of around seven billion people to carry the majority to make big changes in the way we live and manufacture things, starting right now. There are certainly risks in not facing up to the problem of global warming. If we do face up to the risk, that there will be things we can't have any more. Some costs will go up. There will be losers. What each individual needs to do is certainly to change their own life style but also to press their community and government to act – and accept the restrictions which must follow.

Accepting the restrictions is better than the alternative.

Chapter 2: Other Global Research
Looking at the wider environment

The Polar Ice Caps

There is certainly a natural cycle which is not in our favour. There is certainly a significant effect of the human race producing greenhouse gases at an alarming rate. There is no doubt about global warming. There is no doubt that burning fossilised fuel reserves is a key issue. When these are burned, the carbon dioxide produced, commonly referred to as a 'greenhouse gas', or GHG, provides an insulation layer round the earth and that may well be the trigger for melting the icecaps. That melt water changes the flow of ocean waters and the southern oceans rise in temperature, so releasing more carbon dioxide which would otherwise stay in the sea water. Battersby[4] points out that although the orbital changes of the earth round the sun will continue to affect climate in 'natural' cycles, the recurring ice ages are almost certainly gone forever. Conditions now, with the burning of fossilised fuels, have pushed carbon dioxide levels to 380 parts per million and rising. That brings atmospheric conditions increasingly like global conditions ten to fifteen million years ago. Then, the global air was 6°C warmer and sea levels forty metres higher. The reason for raising this argument here is to stress that carbon dioxide levels are affecting our living conditions right here and now and that this book is looking at reversing that trend, taking carbon dioxide out of the atmosphere and putting oxygen back in.

Gavin Menzies was a British Royal Navy submarine commander and, as such, travelled the world and, of course, had access to sea charts as a matter of his daily routing. He became interested in ancient China and amassed an enormous amount of data on the fleets the Chinese Emperor ordered to search the world in the early fifteenth century. The Chinese captains made charts of where they went. Menzies observed that the land had shrunk in many areas. Of course, the accuracy of such charts might well be expected to be less than those made today. However, the differences between the charts made then and those made in the last fifty years are of remarkable consistently shallow seas. You might read his outstanding, well-researched book[5] *1421 – The Year China Discovered the World*. He points out that since 1421, sea levels have risen by "between four and six feet" (1.2 to 1.9 metres). Translating that observation into this text, that could not have been due to human activity all the way back in 1421 because then we were not burning fossilised fuels. Sorry, the rise in sea levels did not occur back then, it has occurred in the last hundred years. Nevertheless, the inference is clear; seas have risen and we have lost large areas of dry land.

---

4.   Battersby, S: *The Great Meltdown*, *New Scientist* May 22nd 2010, pp 33–36

5.   Menzies, G: *1421 – The Year China Discovered the World*, Bantam Books, 2002

There is no doubt whatever that polar ice is melting and there is some evidence that it is doing so at a rate much faster than previously concluded and appears to be accelerating. There are, of course, great difficulties in predicting the rate of melting of ice in the future. So, there is a wide variation is estimates of how this will affect the rise in sea levels.

In the present situation, many of the cities of countries with a coast line were built on the coast because empires were often based on shipping trade. London, for example has a barrage at the mouth of the River Thames which can be lifted if and when high times are expected. It is now on its limit of protection. If the right tide and wind directions come together, it may come over the top. Any time.

Glaciers

Similarly, there is ample evidence that glaciers are melting. The consequences are the same as with polar ice – the local climate will change and the melt will contribute to rising sea levels.

Sea Level and Flooding Risk

We know sea levels are rising and, yet again, the certainty is the uncertainty of how much and how fast. In recent years, there has been a rising level of reports on flooding and flash-flooding from heavy rainfall, at local level. In an article in the *New York Times*,[6] it was reported that engineers from the Netherlands were bidding to assist in the New York area of the USA in flood prevention:

> The Dutch way of thinking is completely different from the US, where disaster relief generally takes precedence over disaster avoidance," said Wim Kuijken, the Dutch government's senior official for overall water control policy. "The US is excellent at disaster management [but] working to avoid disaster is completely different from working after a disaster."

Yet again, what is needed to address risk is *forward thinking* and *prevention* rather than *cure.*

The Oceans

The thing to remember about the oceans is that the amount of water involved is big, very big. Two thirds of the globe is sea and as deep, or deeper, than our land mountains. That water has dissolved carbon dioxide in it which forms a weak acid (carbonic acid). The more carbon dioxide in the atmosphere, the more dissolves in the sea and the stronger the acid which kills off the corals but also alters the whole ecosystem. As the atmosphere warms, slowly it warms up the oceans. As the oceans warm far enough, they give more of their carbon dioxide to the atmosphere than they take in from the atmosphere. That is one of the tipping points. As the atmosphere gets richer in carbon dioxide, it accelerated global warming and here is the trigger for the tipping point.

---

6.   Higgins, A: *Lessons for US from a Flood-Prone Land, New York Times*, November 14th 2012
     https://www.nytimes.com/2012/11/15/world/europe/netherlands-sets-model-of-flood-prevention.html

Exactly when it will occur is not known, except that it is not a point in time, it is an exponential curve which gradually gets away from our potential to control it and hold and/or reverse it. That is just the risk point; wait until we get there and we will have already lost control. However, because of the scale of the oceans, this is not just one of many possible tipping mechanisms, it is the biggest and we have limited knowledge of how it works and how to control it.

This underlines a practical point about the oceans. They are as complex in their movement and flows as the atmosphere but our knowledge of such movements is very limited. Further, our potential to control them is very limited indeed. So, trying to manage the carbon sink in the oceans is certainly something we should research and develop but, in the short run, it is not likely that we will make progress fast enough on a big enough scale.

We know, however, that the centres of origin of tropical storms and hurricanes are moving away from the equator and towards the poles. That movement is gradual but it is there and will probably accelerate. Moreover, the intensity in wind speed and duration is likely to increase. So, the current temperate zones are likely to get both windier and wetter.

Forest Fires

Although forest fires are sometimes started by vandals or man's 'innocent' folly, they are also a natural occurrence. The problem is that the evidence we have indicates such fires are more likely in some vulnerable areas, plus more of them and greater intensity. While forest fires are expected in late summer in place such as North America, such blazes were far worse in late 2018 than previously. More worthy of note is that there were more fires in temperate zones such as the British Isles, Sweden, Iceland and even north of the Arctic Circle.[7]

Species Migration and Loss

There are many documents, some authoritative, indicating profound changes in global wildlife. It is not just climate change (which is certainly driving major shifts and losses) but pollution, loss of habitat and the expansion of human activity that is also a potential tipping point risk. Many changes and losses will never be regained. Desmond Carrington in *The Guardian* quoted a report from the World Wildlife Fund[8] stating that "humanity has wiped out 60% of mammals, birds, fish and reptiles since 1970, leading the world's foremost experts to warn that the annihilation of wildlife is now an emergency that threatens civilisation."

7.   Minnemeyer, S: *Five Graphs Show Just How Unusual This Year's Wildfires Are*, *World Research Institite*, August 20th 2018 https://www.wri.org/blog/2018/08/5-graphs-show-just-how-unusual-years-wildfires-are

8.   Carrington, D: *Humanity Has Wiped Out 60% of Animal Populations Since 1970*, *The Guardian*, October 30th 2018 https://www.theguardian.com/environment/2018/oct/30/humanity-wiped-out-animals-since-1970-major-report-finds

More than 40% of insect species are declining and a third are endangered, the analysis found. The rate of extinction is eight times faster than that of mammals, birds and reptiles. The total mass of insects is falling by a precipitous 2.5% a year, according to the best data available, suggesting they could vanish within a century. "If insect species losses cannot be halted, this will have catastrophic consequences for both the planet's ecosystems and for the survival of mankind," said Francisco Sánchez-Bayo, at the University of Sydney, Australia.[9]

9.   Sanchez-Bayo, F & Wyckhuys, KAG: *Worldwide Decline of the Entomofauna: A Review of Its Drivers*, *Biological Conservation* Volume 232, April 2019, pp8–27

Chapter 3: A Discussion About Pollution

Energy

All life uses energy. Humans use rather a lot of it and, until recently, there were very few moves away from burning fossilised fuels to get it. Of the energy we use, around 45% is for heating. Most of that comes from burning fossilised fuels which, of course, produces carbon dioxide. However, such fuels not only produce that greenhouse gas, but other pollutants, too. Lignite, or 'brown coal' is used in significant amounts by, amongst others, Germany, which has announced plans to shut down all coal fired power stations by 2038. Currently coal provides 40% of the country power and Europe's biggest economy, has already missed its 2020 target to cut carbon emissions. By 2030? The politicians still don't get that the situation we are in is quite urgent. Similarly. China is a bad polluter – hence the infamous smog in Beijing. Next one down in the polluting hierarchy is coal and then the various hydrocarbon oils, natural gas and 'unconventional gas' (shale gas) are fairly clean burns, i.e. they produce carbon dioxide and not much else.

In places such as Canada, '*hydro*' is electricity from hydro-electric generation from dammed waterways. This clearly has a very low pollution output. Nevertheless, all power plants had to be manufactured in the first place, maintained and decommissioned and the energy cost, and possible pollution, from the full life-cycle are sometimes forgotten (see *Section 3*).

Surrogate energy use. There is a quotation, "no man is an island", which must, in absolute terms, be true. We all depend on others for some of our inputs but some more than others. Some, such as a back-woodsman in the American Old West got close to real self-sufficiency but he too relied on others for his gun or ammunition. Genuinely there would be the Australian Aborigines, or the South African Bushmen. At the other extreme, an electric car is made by someone else and powered by electricity made remotely; all the driver does is drive it on a road made by someone else. The driver lives in a clean situation because he/she has exported production and pollution onto someone else.

In farming, we import mineral fertilisers and if we are going to feed the global population, we will continue to use them. As so often remarked on in this text, mineral fertilisers especially the nitrogen, use enormous amounts of energy to manufacture, but also more to package, transport, store and apply. The same remark applies to the manufacture of equipment especially tractors and other engine-on machines such as combines, and all other equipment and the fuel to drive them. In the UK, Massey Fergusson, until around 1985, operated the biggest tractor manufacturing plant in the Western world at Banner Lane in Coventry. Now, very few tractors are manufactured in the UK; we have exported the pollution produced in manufacture but that pollution is still there globally. Governments signing agreements at Paris and Poland to limit their own pollution is laudable – but not if it just means buying things that still cause emissions of carbon dioxide and other pollutants elsewhere.

Pollution

Hydrocarbons and the Carboniferous Era

The Carboniferous Era started 360 million years ago and it took sixty million years to lay down our fossilised fuel reserves. During this period, the proportion of oxygen in the air was less, the proportion of carbon dioxide more, and it was much warmer. That lower proportion of oxygen then and more now may be very important (see *Section 4*). It has taken *less than two hundred years* to burn *more than half* of those reserves.

We have been burning peat and coal for a very long time but in large quantities for only a couple hundred years. Of the hydrocarbon fuels we now burn, lignite, or 'brown coal' is the most polluting producing not only carbon dioxide but also a wide range of other potentially noxious chemical pollutants and particles. Next in order of environmental damage is coal, followed by hydrocarbon oils (diesel and then petrol) and finally, shale gas (the latter is an almost 'clean burn', producing carbon dioxide and not much else).

One of the problems we have is that energy sources, such as wind turbines and solar panels, still use energy to manufacture them and in their whole life management including decommissioning. Those functions are often based on processes using energy from hydrocarbon sources.

CO_2 and Other Greenhouse Gases

It is certainly the case that carbon dioxide, produced by burning hydrocarbons in air, is the biggest worry with respect to global warming. It is also true that despite nations signing up to the *Paris Agreement* on climate change, we are globally behind target on getting emissions of this gas down to signed-up levels.

According to Kelly Levin[10] writing on the World Research Institute website in 2018, the global carbon budget to limit the global temperature rise to 1.5°C will be exceeded in twelve years, i.e. by 2030. What the paper underlines is the need to peak global emissions before 2030 to have a better chance of avoiding the worst climate impacts.

Attention is certainly centred on carbon dioxide and that is certainly logical, provided, that is, that the rest are not forgotten. For example, methane is produced in the stomach of ruminant animals (cattle, wildebeest, antelope and many others) and belched out into the atmosphere. There are many millions of these animals. Methane is a GHG (greenhouse gas) about forty times as potent as carbon dioxide in causing global warming. And there are many more of very damaging pollutants.

Urban Wastes: Obvious and Visible

On the face of it, MSW (municipal solid waste or 'garbage') might not be thought of as significant greenhouse gas issue. However, most of the contents used energy to make them, energy to transport to their point of use, energy to take them away and 'dispose' of them. That energy use is quite likely to have involved burning hydrocarbon fuels. If 'disposal' involved incineration, then consumerism produces even more carbon dioxide – plus other pollutants, too (see *Sections 3 & 4* for an alternative to reuse).

---

10. Levin, K: *According to IPCC, the World Is On Track to Exceed Its 'Carbon Budget' in Twelve Years*, *World Research Institute*, October 7th 2018 https://www.wri.org/blog/2018/10/according-new-ipcc-report-world-track-exceed-its-carbon-budget-12-years

There is no fundamental reason why MSW should not be composted because most of such material, from whatever source, is likely to contain significant proportions of organic carbon and plant nutrients. However, in practice, it is also likely to contain a wide range of pollutants, including heavy metals from the disposal of batteries. Whatever is said about source separation at domestic level, it cannot in practice be trusted. So, separation needs to be done at an industrial level in a Materials Recycling Facility (MRF). Provided the pollutants are at an environmentally acceptable level (whatever that is) (see *Section 5*) such composted material may be used in forestry or to establish tree belts and wind breaks using shredded material and trench reservoirs (see *Chapter 20*).

Urban Wastes: Low Level Insidious

Drug, hormone and antibiotic residues – We have known for over thirty years that hormone residues from women taking the contraceptive pill passed through the body, into the sewage systems, through treatment, into the rivers and then up the food chain via male stickleback fist which produced eggs (the eggs were infertile, of course.) In 2018, some Australian research identified sixty-nine named pharmaceutical drugs in river mud[11] and up the food chain into fish and found doses close to those given to humans. Effects of the doses on wildlife were quoted in the research paper as 'unknown' but it is not logical to assume that there would be little or none. *The New Scientist* quoted[12] the research paper lead, Erinn Richmond, as saying, "if we went to a doctor and said we were on sixty-nine drugs, they would probably be concerned."

Motor and Industrial – Reducing motor and industrial pollution is easier said than done. For example, electric cars are too often seen as less polluting, and they are a bit, only a bit, less. Unless the electricity used to charge the batteries comes from hydro-electric power (as it mainly does in Norway and Canada, for example) it probably has come from burning hydrocarbon oils, which will have produced pollution (albeit at a remote power station, away from the cities). However, as Sean Clark reported in *The Guardian*,[13] "multiple studies have found that electric cars are more efficient, and therefore responsible for less greenhouse gas and other emissions than cars powered solely by internal combustion engines". An EU study based on expected performance in 2020 found that an electric car using electricity generated solely by an oil-fired power station would use only two-thirds of the energy of a petrol car travelling the same distance. There is a real problem in changing the population of existing vehicles to electricity in a time-frame relevant to the targets agreed in Paris and Poland – and preferable better than that.

Industrial pollution from energy use is likely to give us big problems. Certainly, a shift to electrical power will help but, again, the original source of the power may be burning hydrocarbon fuels. Until there is a complete shift to renewable electricity, it will remain a problem.

---

11. Richmond E et al: *A Diverse Suite of Pharmaceuticals Contaminates Stream and Riparian Food Webs*, *Nature Communications*, November 6th 2018 https://www.nature.com/articles/s41467-018-06822-w

12. *Medical Waste Chemicals Found in Beetles*, *New Scientist*, November 10th 2018, p7

13. Clarke, S: *How Green Are Electric Cars? The Guardian*, December 21st 2017

Housing – In the UK, 45% of all energy use is to heat houses. The problem with building a zero-energy consumption house, which is quite possible, it demands significant energy spent on insulation and building solar panels and other energy sources such as groundwater heat pumps. The real problem is not if we have the technology to do this (we do!), it is in changing the housing stock fast enough.[14]

One bit of good news, for a change. We are slowly getting around to building a roof from solar panels. Not, that is, building a roof and retro-fitting solar panels on top (i.e doubling the cost of the roof) but, instead, dispensing with the conventional roof and building of solar panels only. If every roof of every building globally were so constructed, and we had the batteries to hold power in store, then we would not have a housing energy problem.

Whatever you believe about global warming, it will bring major, fundamental changes in the way agriculture is run and, in particular, the way water is managed. So how abrupt will the rate of climate change turn out to be and what can be done to plan production to accommodate and exploit these changes?

Jet Planes and Contrails – Few who talk about climate change will know much about, or have even heard of, 'global dimming'. When fossil fuels are burned they produce carbon dioxide, a *greenhouse gas*. Increasing amounts in our atmosphere allow more sunlight through to the surface of the earth, and air and surface temperatures rise. Most observers, including those in the USA (despite President Trump's views), accept that this is happening although there is discussion about how fast. Clearly, whatever the speed of temperature rise, it will affect all of us. It is doing so already. Global dimming counter-acts that effect. One of the parallel effects of burning fossil fuels and of industrial activity is that dust, soot and 'dirty' chemicals are produced into the air. These have many effects, most of which are seen as harmful. However, it is also true that they tend to settle and concentrate on clouds in the sky, thus turning the clouds into mirrors which reflect sunlight back into space. This reflection of sunlight by clouds is called '*global dimming*'. So pollution has, to some (but limited) extent, been balancing out with the dimming effect of dirty clouds, thus compensating for greenhouse gasses. How good is this balance? Well, one of the more interesting results of the 9-11 terrorist attacks in the USA is that most aircraft in the world stopped flying for three days. This allowed what is called a 'natural experiment' i.e. something happened on its own and allowed observations to be made. It is also true that the contrails, or 'jet steams', the trails left in the sky by the exhaust of the engines of jet planes, also has a significant global dimming effect. When aviation fuel is burned in a jet engine, it produces carbon dioxide and water. At altitude, it is very cold and that water freezes. What we can see in the contrails is billions of ice crystals. These reflect sunlight back into space and causes the dimming which reduces the light and the warming effect of the sunlight.

So, at best, there is a mixed blessing. Unfortunately, the situation is not sustainable because the carbon dioxide emitted is by far the bigger effect.

Water – For many, especially in warmer areas of the globe, water is likely to become as important as petroleum fuels. If you have water and can manage it in the face of whatever these climate changes bring, then you will be better placed than most in maintaining life

---

14. Le Page, M: *Global Cooling Starts At Home, New Scientist,* November 17th 2018

and growing business.[15] The building of better water management is essential at every level, including international and local politics, industrial planning, and at farm level. This means managing the water supply chain. Our society's survival depends on our ability to manage water better and better. We need to examine losses and efficiency of use at every step. Perhaps this is no more evident than in Israel where high-tech farming produces good crops from recovered desert but there is a threat of water running out; the 'water clock' in Israel is ticking.[16] A simple check-list[17] for immediate action might be:

- *Aquifers:* Where are they? Can we find more? What are the risks involved in their long-term use? How can they be filled and used efficiently? How can their pollution and long-term damage be avoided?

- *Dams and Reservoirs:* Do we need to build more? When? Can they be built at farm level? Can dam building be integrated with energy production from hydro-electric schemes?

- *Water Distribution:* Distribution can, and often does, entail losses of up to 50% of what was conserved in the reservoirs. What exactly are these distribution losses? Can they be eliminated, reduced or exploited for some use?

- *Air Management – Tree Belts:* Tree belts can change local climate and reduce soil erosion risks.

- *Top Soil Reservoirs:* Raising soil organic matter can increase its retention rate to hundreds, sometimes thousands, of tonnes of water per hectare from the rainy season into crop growth and harvest. Cultivation oxidises that organic matter. Raising organic matter can easily be done using composted urban wastes.

- *Irrigation:* Surface irrigation using hand-dug channels is cheap but extremely wasteful in water use. At the other extreme, a 'spaghetti' line to each plant is efficient but expensive. Better still, use *top soil reservoirs* (see *Chapter 21*). Climate change will push the balance in the direction of more expensive systems in order to use water better.

- *Cultivation Management and Zero Tillage:* Cultivations turn moist soil upwards to be dried in the sun; this oxidises the organic matter. Zero tillage systems dramatically reduce these effects. So cultivations and cropping may have to change to take advantage of these facts.

- *Crop Choice:* Some crops can cope with drought stress better than others. Similarly, some varieties within a crop species are better in this respect than others.

Water is a political issue. We can view water as maybe the most important of all our assets which we must manage better, using all our existing knowledge in addition to a major research effort to do better. Man can live without oil but not without water.

---

15.   United Nations Statistics

16.   Swinburne, Z: *The Water is Running Out in Gaza: Humanitarian Catastrophe Looms as Territory's Only Aquifer Fails, The Independent (UK)*, May 2015

17.   Butterworth, B: *Sewage Solution for Low-Water Farming, Far Eastern Agriculture*, September/October 1997

Water is already more valuable per litre than oil in many parts of the world. As the population increases and the earth warms up, that is going to be more widespread and more acute. It is not unreasonable to suggest that water is already more important than energy in terms of work politics and power. It is not so dramatic as oil but it, or lack of it, changes everything. Nowhere is the critical inter-relationship between water and energy more evident than in the Asia-Pacific region, home to 61% of the world's people, and with its population expected to reach five billion by 2050. The Asian Development Bank (ADB) forecasts a massive rise in energy consumption in the Asia-Pacific region: from barely a third of global consumption to 51–56% by 2035. Similarly, *The Independent* newspaper in the UK reports:

> With 90 to 95 per cent of the territory's only aquifer contaminated by sewage, chemicals and seawater, neighbourhood desalination facilities and their public taps are a lifesaver for some of Gaza's 1.6 million residents. But these small-scale projects provide water for only about twenty per cent of the population, forcing many more residents in the impoverished territory to buy bottled water at a premium. The UN estimates that more than eighty per cent of Gazans buy their drinking water. "Families are paying as much as a third of their household income for water," said June Kunugi, a special representative of the UN children's fund Unicef.[18]

Agricultural Pollution: Methane
The global population of domesticated cattle is around one billion and they produce around 75 to 80 Tg of methane per annum. Methane is around forty times more potent as a GHG than carbon dioxide. According to Eckard et al,[19] global agriculture produces approximately 10%–12% of total global human-sourced greenhouse gas emissions, contributing around 50% of all methane (CH4) and 60% of all nitrous oxide (N2O). Apart from their significant contribution to greenhouse gas emissions, the energy lost as methane and total N losses are two of the most significant inefficiencies remaining in ruminant production systems. Note that, as this text will show, the nitrogen losses can be substantially avoided and the energy use reduced.

Agricultural Pollution: Manures
In terms of pollution risk, animal manures contain two fundamental components; soluble and insoluble. The soluble ones contain many crop nutrients, including nitrates, which can be held on high organic matter soils (see *Section 4*) but some will be leached out in rain or irrigation. The insoluble components will help build up the soil organic matter (and that will help hold future additions of soluble nutrients). The soil organic matter is where the soil fungi *mycorrhiza* live, multiply and feed the visible plants. The management of this biosphere is a key component in managing the bio-active carbon sink (again, see *Section 4*).

18.   State of Palestine WASH News update, *UNICEF*

19.   Ekard, RJ et al: *Options for the Abatement of Methane and Nitrous Oxide from Ruminant Production: A Review*, *Livestock Science*, Volume 130, Issues 1–3, May 2010, pp47–

Agricultural Pollution: Residues from Pesticides, Sprays & Mineral Fertilisers

There can be no doubt that agricultural chemical residues have, in many parts of the world, been a serious and sometimes under-reported risk – and continue to be so. The same is true on mineral and manufactured fertilisers, especially nitrogen. Even in the EU where the application of nitrogen is significantly restricted by law, especially out of the growing season, there is a loss to groundwater of in the region of thirty to 55% depending on soil type, organic matter and rainfall/irrigation rate (once again, see *Section 4* for reducing and possibly eliminating these losses).

There is an aspect of this potential pollution which is important and it is about the balance of risks. Food produced in the developed west has, generally speaking, a low level of residual chemicals in crops and soils and, in the EU, and UK in particular, because of regulation and environmental policing, possibly the safest food in the world. However, there is still a risk. It is necessary to balance this risk against the alternative. 'Organic' farming on the face of it is very attractive. In the hands of a good farmer, it can work reasonably well, provided that is, all the neighbours are fighting off pests, mostly by using the effective tool of chemical sprays. Just suppose, for the sake of balance of risks, that a single hectare (ha) of crop failed. Such failure are dramatically more likely with organic farming. According to World Bank figures[20] the world average requirement for arable land to support one person is 0.194 ha. So the answer to the question of one hectare crop failure? On that World Bank figure, one ha of failed crop would cost five lives. However, it really depends on where you live; if you live in the UK and in a good farming area, and that one ha fails, you will have lost maybe more than ten tonnes of wheat. And that, at around 45 kg per person per year, means that, somewhere in the world, 220 people die of hunger. It is possible to argue at length about the exact figures but it is true that, generally speaking, the sensible use of agricultural pesticide chemicals has saved lives and we could not feed the world population without them. The truth is that sprays such as the very safe glyphosate ('*Roundup*') has saved hundreds of millions of lives. The choice is simple – we either use the crop protection chemicals (with increased policing to see that people obey the rules) or shrink the global population to around half the current level, i.e. somewhere around 3.5 billion. Put bluntly, if you favour the universal use of organic farming, would you like to volunteer to be one of those who die now? However, there is a conundrum here. Recent reports[21] indicate that we may have already lost 40% of insect species, at least in part because of the use of agricultural sprays, and there are serious threats to pollination of crops, food production and the dynamic balance of the global ecosystem.

Global Carbon: Atmospheric

This has been present as carbon dioxide and it has been there ever since life as we know it began. Most of life (but not all) is based on breathing oxygen and, in the production of energy, the body produces carbon dioxide. Further, carbon dioxide content of the atmosphere has been as high as 4000 ppm if we go back 500 million years and has fluctuated wildly since – and so has the temperature of the atmosphere. For maybe

---

20.   Arable Land – Hectares per Person, *The World Bank*

21.   Sanchez-Bayo, F & Wyckhuys, KAG: *Worldwide Decline of the Entomofauna: A Review of Its Drivers*, *Biological Conservation* Volume 232, April 2019, pp8–27

ten thousand years up to around 1750 the figure was fairly stable at around 289 ppm. However, the industrial revolution began to burn coal and we have escalated the burning of all fossilised fuels ever since and carbon dioxide in the atmosphere rose to around 340 ppm by 1989, and to 405 ppm by 2018. That rise put a very significant risk on the survival of human civilisation as we know it.

That figure of 4000 ppm 500 million years ago has a very interesting consequence. Plants which evolved then did so under the conditions at that time and the process of photosynthesis, which used energy from sunlight to take carbon dioxide out of the atmosphere to make sugars and hence cellulose (and give us the oxygen back) was evolved under that concentration of carbon dioxide. Unfortunately, that mechanism does not work so well under the carbon dioxide concentration in our conditions now. This may give us a route to make plants much more efficient. Some potential breakthrough research on this mechanism is discussed in *Section 4* below.

Global Carbon: Oceanic
The carbon dioxide content of the oceans is one of the big "if's" – we do not really know how dangerous this is in factual numbers terms but it does look pretty resinous. Historically, the oceans have absorbed carbon dioxide out of the atmosphere by two routes; one just by absorbing it into the water (which forms the mild acid – carbonic acid) and secondly, sea plants (including the phytoplankton) do what dryland plants do – take carbon dioxide out of the air/ water and use the energy in sunlight to make sugars – and give back oxygen. That is fine – BUT as the atmosphere warms up, so does the ocean and as that happens enough, it begins to give up its dissolved carbon dioxide, which in turn warms up the atmosphere, which in turn causes the ocean to give up more carbon dioxide which... This is probably the biggest of the tipping point risks. Over 90% of the global carbon dioxide is stored in the oceans. If we get to the tipping point, the rate of warming will dramatically accelerate. That is a very real risk.

Global Carbon: Forests
Forest soils hold enormous amounts of organic carbon. Oxidise the carbon and the result is carbon dioxide released into the atmosphere. Burning forest in air and removes oxygen and produces carbon dioxide. The process of cultivating the soil so released for farming does, in turn, produces another significant amount of the GHG and does so every year the ground is cultivated.

Burning down tropical jungle, as is happening in Brazil, Indonesia and other places, may justified by short term economic drivers – but environmentally is disastrous.

Global Carbon: Soils
Soils globally hold an enormous amount of carbon. While attention is often focused on carbon locked up in trees, in fact, most of the terrestrial carbon lies in the soil. Below ground carbon includes an array of sources such as the root systems of trees and soil organic matter. By managing the world's land more sustainably, such as by protecting forests and investing in reforestation, we could make a significant contribution to emissions reductions necessary to limit the global rise in temperature to 2°C by 2030. So, inescapably, recycling urban, organic carbon wastes to farm land by composting is a necessary development. Only farmers can do this.

Cultivating the soil increases oxidation rates. As a guide, conventional cultivation based on the plough and maybe two or three more passes with cultivators to produce a seedbed, will oxidise maybe 35% of the organic carbon each year. Moving to zero till seeding, often called '*direct drilling*' in the UK, will reduce that oxidation to around 10%.

So, soils are potentially a really useful tool in arresting global warming. Again, there is a problem; according to one report[22] the Earth has lost 33% of its arable farm land in the last forty years. Before looking at population, there is something that is worth remembering about what people stand on – Land; and they really have stopped making it. Indeed, we appear to have lost a lot already[23] if sea levels rise as predicted, we are going to have significantly less of it. What is more, the land we will lose, however much it turns out to be, is, in the main, the best, most productive farm land. People need food and we have more people and less land – and we are losing it to rising sea levels, erosion, desertification and building, every minute of every day.

Logically, as population grows, expansion of food production is going to fall, and maybe go into reverse. Here, again, is a risk and potential influence of the population tipping point.

Population and Conclusion

The United Nations, with considerable resources, argues that population will level off. Professor Hans Rosling has become expert in presenting statistics on population growth to a wide audience. He argues, with significant evidence, that there is a rapid, global trend to families having only two children.[24] Globally, he argues, this is a very positive situation and he is hopeful that the world will avoid disaster. Well, maybe?

The truth is that at present, the population is still expanding and people exacerbate all the above problems of carbon dioxide production, demand for more food, and pollution. All of these are becoming greater. Whether we like to admit it or not, birth rates in wealthy countries are falling but death rates are falling faster. This subject will be covered in more detail in *Section 2*.

---

22.   Milman, O: *Earth Has Lost a Third of Arable Land in Past forty Years, Scientists Say*, *The Guardian*, December 2 2015

23.   Menzies, G: *1421 – The Year China Discovered the World*, Bantam Books, 2002

24.   Rosling, H: *Global Population Growth Box by Box*, *TED*

Section 2

Population and the Failure of Antibiotics & Vaccines

Chapter 4: Population, People and Problems

Population Curves and Dynamic Balance

All living creatures migrate. They do it to try to find better conditions. They always have and always will. Early man emigrated outwards from Africa. For centuries, and in current times, the wildebeest in Africa migrate seasonally. In a cold winter when bees run out of honey they may go on a hunger swarm in a last ditch attempt to survive. It is in the nature of the survival instinct. In the last 200 years of so, while the human race may not have appeared to those comfortable people living in the West and looking back, to have been moving about much, they have. There was massive forced movement during the slave-trading years, largely from Africa to the USA. Recently, there has been much publicised mass movement of millions out of Syria, Libya and North Africa generally, to Europe. As climate change takes tighter hold of some areas, probably parts of Africa and Asia, then 'hunger' migration will increase. Population growth will just increase the pressure on resources and encourage that survival instinct to go look over the hedge in a search for a better life.

Alternatively, where conditions are favourable, with most species, birth rates tend to increase. However, some well-informed commentators argue that this may not always be the case with humans.[25]

If you look at almost any report on current world population figures, it will look something like the first figure overleaf:

We can be fairly confident of the date and, for the numbers, there are some reasonably reliable figures on world population. For the logic of this argument, it does not really matter whether the population at the time of writing is taken as seven billion or even ten. By the time you read this, it may be 7.5 or 8, or beyond.

The second figure shows a typical, idealised, stable population curve which levels out at a 'comfortable' maximum and stays there.

In theory, population curves such as this do exist. However, there are two points to remember. The first is that it is never a smooth climb up the curve; there are always occillations up and down, as with individual births and deaths but also more abruptly and with larger oscillations because of changing conditions. The second is that, because of these changing conditions in a real world, there is not a smooth plateau eitheer, it is a dynamic balance, always changing and trying to create a stable situation in an ever-changing environment.

Everyone assumes that this sigmoid curve which levels off is what it will be like for the human race and then argues about at what point it will level out, or assumes that we can keep on going. Only an idiot would argue that there is no limit; ultimately, there has to be a point where, however efficient we are at meagre living, at efficiency of resource use, the population will outstrip the available resources and level off.

---

25.   Rosling, H: *Global Population Growth Box by Box*, *TED*

A TYPICAL POPULATION CURVE

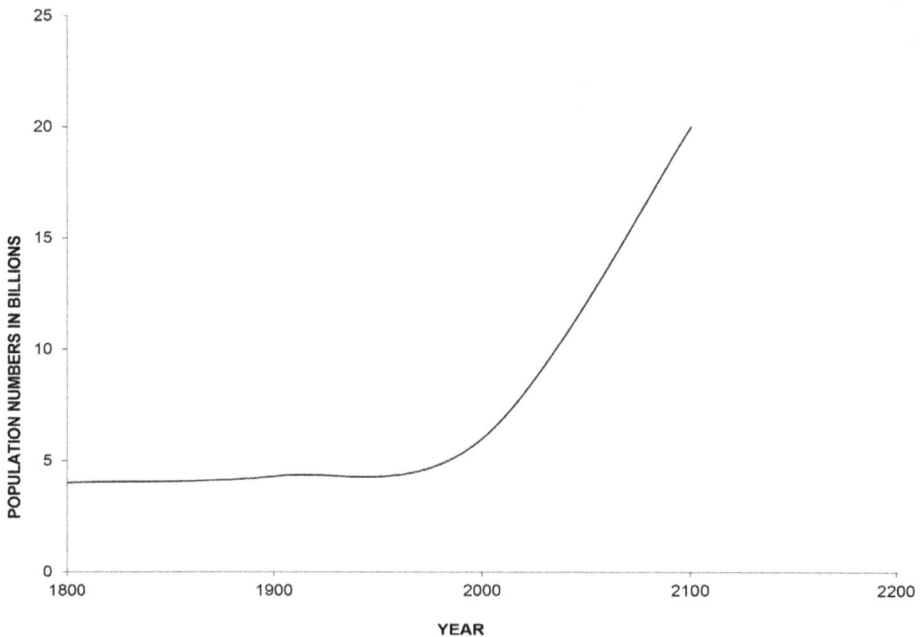

Fig 4.1: World Population To Date

A TYPICAL, IDEALISED POPULATION CURVE

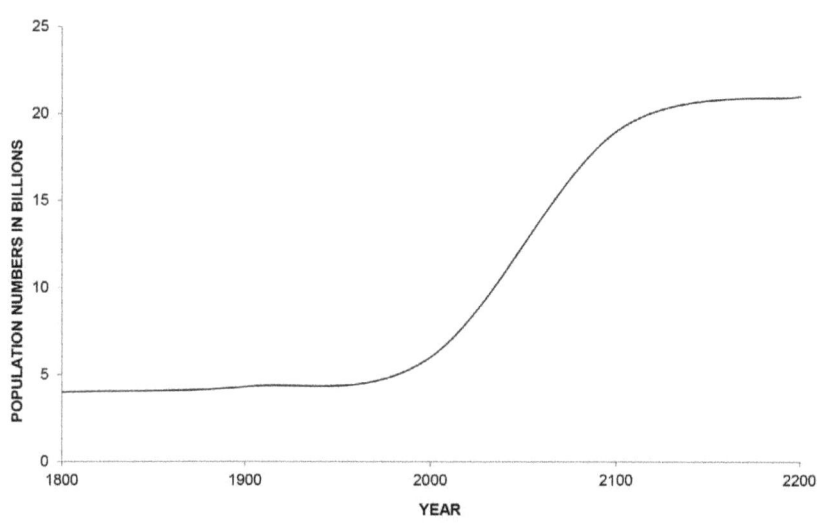

Fig 4.2: A Typical, Idealised Population Curve

Most reasonable discussion appears to argue that it will level off at fifteen to twenty billion sometime in the next fifty or one hundred years. The truth is that such estimates have a number of weaknesses. Nevertheless, it is a nice picture and it makes us feel comfortable, partly because we feel we don't have to face the process of how it levels off, or whether there will be some other happening at that point. Do populations level off smoothly to a stable situation and are any of the figures of fifteen or twenty billion a reasonable guide? How do we know that it will be one hundred years off? Might it be fifty years? Might it be tomorrow morning? How will it level off? What will be the process?

A key to the answers to these questions can be seen in William Stanton's thorough study of world population curves.[26] Nearly every country or regional population in the world has hit this exponential rise. Generally, birth rates have stayed high and death rates fallen. It is certainly true that many developed economies eventually show a stabilising population because birth rates fall. Indeed, many argue that this will solve the problem in that every region of the world will follow the same pattern and that world population will stabilise and the application of medicine and economic wealth will proceed to levels which are, at least in part, predictable levels. There is plenty of evidence for this in the academic and government literature on population.

The figure opposite gives the United Nations latest predictions at the time of writing in 2018. The general prediction is that the level of growth in population is levelling off but note that it is not quite so in most of the curves in the graph. Population can only fall if a couple have less than two children.

Professor Hans Rosling has become expert in presenting statistics on population growth to a wide audience. He argues, with significant evidence, that there is a rapid, global trend to families having only two children.[27] Globally, he argues, this is a very positive situation and he is hopeful that the world will avoid disaster. Will that trend be fast enough? Well, there will certainly be some adjustment and pain. The idea that *every* region and country will ever do this does, logically, seem very unlikely. We do have the technology, in place right now, to actually feed everyone in the whole world but we don't for a complexity of reasons which are political, to do with military power, and logistical difficulties in equalising distribution of resources. The latter reason will always apply simply because of the protective instincts of the 'have' nations, versus the 'don't have' ones. Similarly, there will not be an evening up, or rather down, to stability, of population growth in all nations; it is not the nature of things. However, suppose for the sake of argument, that there comes that point of stability in world population as in figure 4.3. What will it look like and will it stay like that? How will it affect greenhouse gas production?

Military War: There is no doubt that wars kill people and, in some cases, a great many people. However, when the war is over there is almost always a compensatory growth back in population numbers. Where history shows that wealthy nations have a weakness is in guerrilla warfare or, weaker still, in 'illegal immigration'. So, conflicts do affect population numbers but this does not affect long term numbers. What does often happen is war stimulated migration.

26.   Stanton, W: *The Rapid Growth of Human Populations 1750-2000*, *Multi-Science Publishing Company Ltd*, 2003

27.   Rosling, H: *Global Population Growth Box by Box*, *TED*

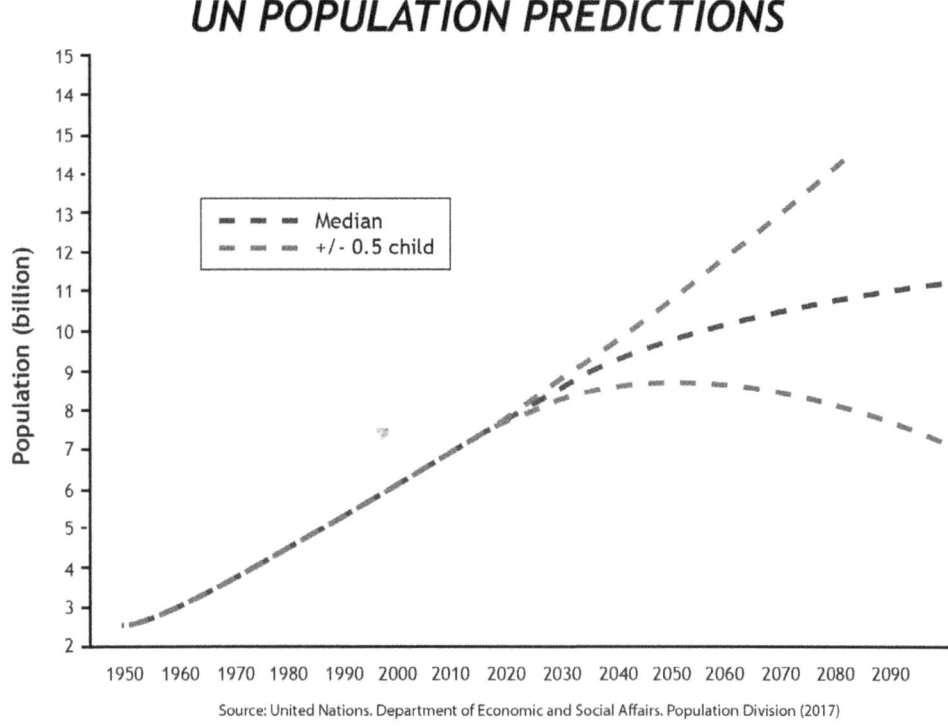

Source: United Nations. Department of Economic and Social Affairs. Population Division (2017)

Fig 4.3: UN Predictions on Global Population Changes

Large countries have large borders. Large borders are difficult to police. It is therefore inescapable to conclude that there will always be pressures to even up apparent wealth. It will go on for decades, centuries, millennia.

Energy Use: If Professor Rosling is right, and he certainly is at least in part, and global population stabilises before global disaster occurs, then the main reason for the professor's logic is a wealth-driver, i.e. that families want to give their kids more and have a 'better' life and retirement. It is inevitably true that means consumption of energy per head goes up and does so dramatically.

All people, even those actually dying, use energy. Wealthy people use a lot of energy, even if they live in a warm climate and don't use a motor car or fly in an aeroplane. High density populations use more energy per person because of the energy cost of logistics to move food and goods, sewage and water, and also because of the increasing demands of health needs and disease control. All our plastics, metals and just about everything we use for the sustenance of life is energy dependent and most of that energy comes from burning fossilised fuels. Even the populations of the developed countries of the world surfing the net is dramatically increasing energy consumption.

There is no escaping it, survival demands energy. Currently, the human race is geared up to using fossilised fuels to find that energy.

Does population collapse actually occur? One of the most interesting curves Stanton[28] draws attention to is that for the British Isles over the last thousand years or so.

There have been catastrophic collapses of population more than once in British history; three from the Black Death alone. This one disease, in a series of pandemics, spread over a comparatively short number of years, more than halved the population of Europe. There are some who take the view that HIV will/is, doing the same in Africa. Will HIV be controlled? Well, in terms of finding a treatment, it has been cracked and at an enormous cost financially and in lives. How many will die before it is eradicated? The truth is that we live with HIV and there is a partial lid on it. The point is that catastrophic falls in population do occur. They occur because population pressure pushes the available resources to the limit of survival and then there is an adjustment.

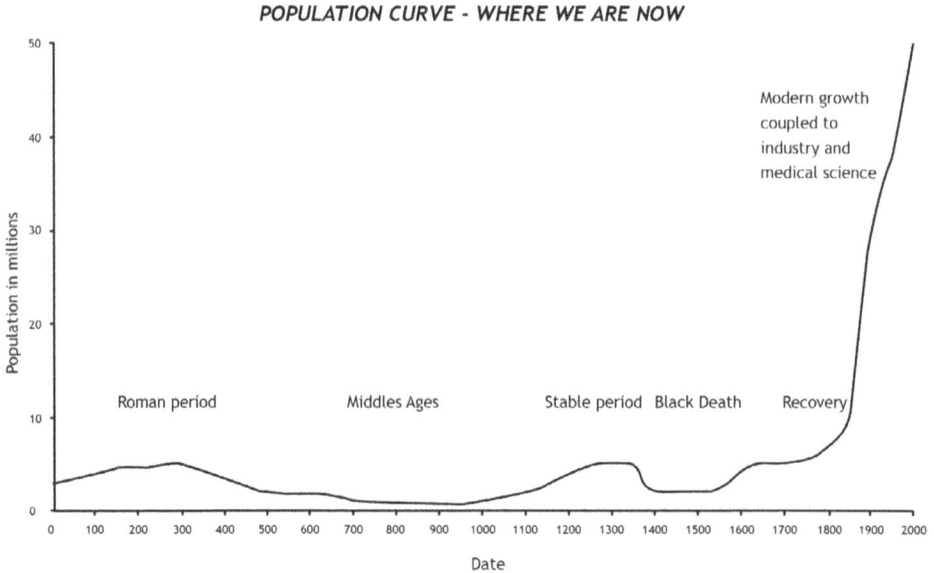

Fig 4.4: History of the Population of the British Isles

This *always* happens; it is the nature of population.

Human populations are different, of course, as we exercise a much greater control over environment than any other species (well, we think we do). We are capable of pushing frontiers back. All of this is true and we will keep pushing the frontiers back. Is there a limit? No, but as population grows, then catastrophic collapse becomes an increasingly likely risk. There are many of these risks but there are two which are worth further thought in order to demonstrate that this is a real problem.

Firstly, population density has a remarkable consistent effect on the social habits of every species. Factory farming is a case in point. Put chickens too close together and they peck the weaker ones to death. Put too many pigs together and they become

---

28.   Stanton, W: *The Rapid Growth of Human Populations 1750-2000, Multi-Science Publishing Company Ltd*, 2003

cannibals. There is no reason to believe that the human species is different. Indeed, there is plenty of evidence to confirm that we are exactly the same. People in cities become more aggressive and self-protective, crime rates go up in unstable communities; it is inevitably the nature of survival.

Secondly, as population increases in numbers and the available space per individual declines (as in growing cities), then it becomes more vulnerable to disease. As population densities rise, then disease transmits from one person to another and the likelihood of mutations in diseases increases, resources become stretched, pollution increases, the speed of infection passing between individuals' increases and the progress to catastrophic collapse increases.

Logically, the likelihood of catastrophic collapse of global human population is unavoidable. There is, of course, argument about what the truly sustainable level of population will be. Suppose, for example, it is in the region of 3.5 billion but population gets to ten billion before disease causes catastrophic failure. Then collapse will overshoot that figure and oscillate until it stabilises and then, over the next thousand years or so, the whole process of population growth will start again – we hope. Remember, the dinosaurs did not make it (To be fair, that population may have/probably did collapse because of a massive meteorite impact on the earth but, nevertheless, it is wise to remember that this is a complex universe and we need to keep our wits about us).

THE CATASTROPHIC POPULATION CURVE

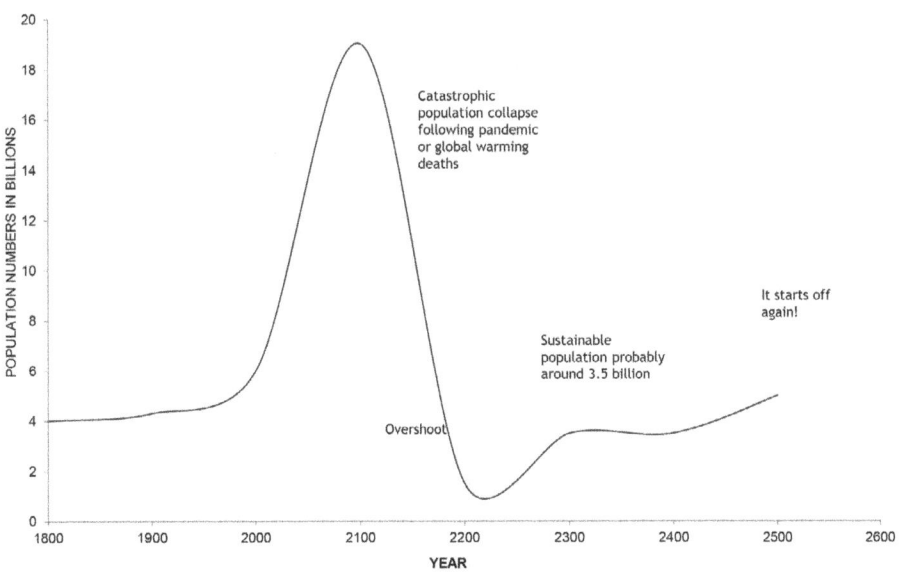

Fig 4.5: The Catastrophic Population Curve

From time to time, really eminent population experts with enormous experience and not given to sensationalist approaches to judgements have predicted this sort of catastrophic collapse. Logically, it could happen and there are good reasons to expect it as a real danger. The most likely precipitating factor is not a disease like ebola, horrifying as it is, as infectious as it undoubtedly is. Ebola, however, is infectious by contact with, for example, infected mucous or blood. In theory at least, that means that isolation methods could control it. The zika virus is one stage worse because it affect the fecundity of the next generation and transmission is by the mosquito which, in theory could be controlled. Isolation or control gets one stage again worse and be much more difficult with air-borne viruses such as influenza. If there was a swine 'flue, an avian 'flue and a couple more, all of which mutated *at the same time* to extreme levels of fatality, control would be very difficult. So, yes, it could happen. However, the human race is incredibly innovative at survival. In the decade up to today's date (2019), vaccines with multi-strain resistance which could cope with *all* viruses of a type such as all 'flu viruses, by a single inoculation were deemed possible and were actively under development.[29]

Successful development of vaccines which could give long term protection against a group of viruses, such as influenza, would be a breakthrough of enormous significance in terms of human population survival and control.

Despite this hopefulness, there is another potential outcome curve. Fig 4.6 opposite shows an un-stable plateau. There will always be new diseases, new mutations of old ones, new limits on local resources, new impacts of climate change (including floods, fires and temperature extremes) and entirely new risks. Not only is it logical to think that this is a likely outcome, we actually know that this will be the outcome. How do we know? Because it is already happening and has been so for many years. Consider much of Africa, right now. The evidence is that we are approaching a plateau but not a smooth one. It is just that the natural process has become more extreme and therefore oscillating.

Summary of Population Growth
As discussed in the very first pages of this book, world agriculture uses a very significant proportion of global energy consumption. It is worth going back to Al Gore's summary of where we are. Gore[30] did not get his Nobel Prize as a trivial passing remark by a group of uninformed idiots. This is serious stuff beyond reasonable objection; there is a serious problem here. There is no doubt that the rise of human populations is occurring. There is no doubt about the increase in the use of energy from all sources. There is no doubt about the increase in greenhouse gases. While there are many different greenhouse effects and many gases worse in their warming effect than carbon dioxide, the fact is that carbon dioxide emissions from burning fossilised fuels is a major issue; some would argue _the_ major issue and _the_ most urgent.

29.   Impagliazzo et al: *A Stable Trimeric Influenza Hemagglutinin Stem as a Broadly Protective Immunogen*, _PubMed_, August 24th 2015

30.   Gore, A: *An Uncomfortable Truth*, DVD, _Paramount Pictures_, 2006

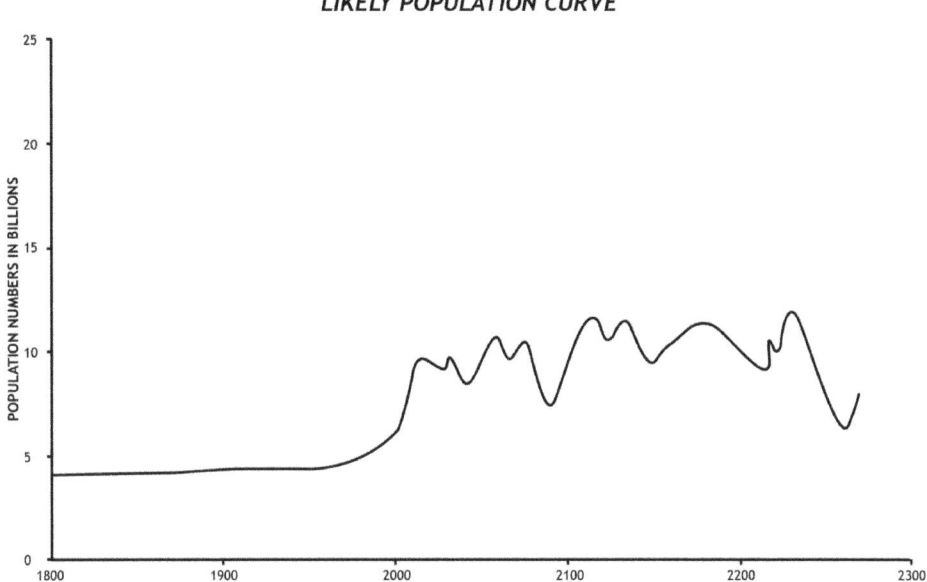

Fig 4.6: An Unstable Population Curve

Conclusion #1: *Population Increase*

Increasing population results in increasing carbon dioxide production and global warming. Whether we like to face it or not, population growth is an issue. We ignore it at our peril. It is nothing whatsoever to do with race, it is to do with numbers.

However efficient and sensible we get in the use of energy, the fact is that humans use energy. Every extra human individual makes a contribution to greenhouse gas production and global warming. It may be possible to have a large population and avoid that, but it appears logically to be unlikely, certainly in the timeframe that global warming is likely to be catastrophic.

So, feeding an expanding population is an issue. That food production implies increased energy use. However, part of the function of this text is to examine ways of producing more food without increasing energy consumption. In the short run, every new human means we burn a little more fossilised fuel and produce a little bit more carbon dioxide.

Conclusion #2: *Postponing Death*

Logically, all the compassion, all the charitable care, all the medical science, they never save lives, *not ever*. What they do is postpone death and, hopefully, make it less uncomfortable. Therefore, the exponential rise of population seems more likely than not and catastrophic collapse (probably due to a pandemic of several mutated virus disease occurring at one time) an inevitable consequence. Fortunately, birth rates are falling but the human race will not survive unless birth rates fall to less than two per family, and very quickly.

There is an outstanding question; if we lack the global political will to stop unfettered population growth, as seems likely, will there actually be a catastrophic collapse and will this, then, solve the issue of humans' contribution to global warming? Would such a collapse stop a catastrophic rise in global temperatures? Probably not; it would be too late. If the human race takes enough action, soon enough, then there will be a plateau in population numbers, but it will be a dynamic balance with significant oscillation in the 'plateau'.

Clearly there has been, and still is, an enormous amount of expertise and work going into population studies and there is still a somewhat hazy prediction on what will happen. On balance, there is good reason to believe that the rate of rise will certainly fall in the second half of this century. As the numbers rise, then some influencing factors will become more important, and some less so. For example, it is likely that disease in some parts of the world will help reduce the total numbers. In some areas, military action will precipitate death and/or famine. In others, shifts in global temperatures and other weather factors will become critical to survival. Repeated reports from the United Nations suggest that just as there were around two billion children in the world population in the year 2000, there will be the same number in 2100 and that total population will level off at around eleven billion. So, together with the trend to smaller families, there is at least some reason to believe that population will level off but it will be a limit under constant pressure with under-lying pressures on human health. So, yes, the human race may well survive but it will not be by accident; there is a constant need to keep pouring resources into health research.

Population and Global Warming.
So, where is this population discussion taking us and what has it to do with reversing global warming? Firstly, population numbers are a real issue. In a relatively recent (i.e. within the 21st century) debate in the UK parliament, David Kidney MP raised this issue with respect to the UK population. He was quoted by Martin Livermore[31] in a piece for the Adam Smith Institute and he also quoted the *Optimum Population Trust* as saying "UK should be limited to thirty million" (at the time, the UK population was approaching seventy million). However, there was a public outcry and the subject was dropped. Further, *The Global Footprint Network* tells us that we (the world) are currently using the resources provided by 1.3 planets. However, Kidney went on to say "It's important, in my view, to keep a sense of proportion and to look at population size in its context alongside other considerations, including developing the means to provide a larger world population with the water, food and the energy it will need."

Disease, Antibiotics and Pandemics
It may be worth looking out for the signs of a critical situation. Pandemics do occur. Although diseases like ebola are frightening and devastating in countries where the social and government infrastructure are weak or non-existent, they are relatively easily isolated in developed countries and that means they can be controlled. However, when a virus is airborne, it is easily transmitted and it is much more difficult to isolate those carrying the disease. As population density increases, then transmission is more likely and, as transmission occurs, mutation is more likely. Mutation may make the virus less virulent – or more. If it is more, then fatality rate increases.

---

31. Kidney, D, MP: *Global Population Growth*, <u>*Debate in British Parliament*</u>, February 4th 2009

If and when there are several viruses of this type, all mutation to be more fatal at the same time, then the problem may get out of hand. Recognising this in advance is an important part of control. The most likely viruses to do this are the flu viruses and two of significant global threat during the first decade were commonly labelled *avian* and *swine*. If several different and virulent strains of these viruses arrived on the global stage at the same time, then the human race would have a potentially serious problem.

This, then, raises the question of what can be done.

In tackling disease, there is one important fact of life that needs constant attention and care. Consider the size of a bacterium, and even smaller, a virus. Most antibiotics will not tackle viruses, so far, so just consider bacterial diseases. Consider maybe a few billion wandering round a body or in the wider environment. The idea that it is common practice to rid a body, worse still an area of environment outside the body, of 100%, of that organism, really 100%, not one bacterium body left, is not common sense. There is always a hiding place and, therefore, the battle is never ending. The body may manage to establish a dynamic balance (i.e. seen as 'healthy') but is it still a constant battle. Now, if that is at least commonly true, even sometimes true, that means that the use of antibiotics does not kill 100%, not every single one, of the pathogenic bacterial bodies. That means some remain alive. Some may have escaped any attack by the antibiotic. Some will have been attacked but have survived, i.e. they are resistant to attack. The process is *always* selecting out resistant strains of the disease.

The therapeutic use of antibiotics, i.e. in the treatment of identified, visible disease, is widespread and has saved countless lives and suffering. Nevertheless, that process will certainly and continually select out more resistant strains of every disease so treated. So we will need to keep developing new antibiotics. However, the process of selecting out resistant strains of disease is much more rigorous than that. The prophylactic use of antibiotics, in a small dose to prevent the disease, often done for long periods, is globally widespread. This, inevitably selects resistant pathogens more actively. With livestock, both in agriculture and by veterinary surgeons in pet keeping, such prophylactic use is very widespread and often for the whole life of a farm animal.

This inevitably means a constant battle to develop new antibiotics. If we do not do this, the pandemic is sooner rather than later. The same argument applies to vaccines and the various influenzas are an example of the constant mutation of these viruses and the constant up-dating of the vaccines involved, administered on a global scale.

Vaccines. The answer of course, is a 'flu-universal' vaccine. There is, fortunately (from a health point of view) a potential answer to the flu pandemic and that is in a multi-stain vaccine. There have been significant progress, initially reported in 2015, of discovery of the basic mechanism to achieve this. Conventional flu vaccines target the 'head' of a molecule called hemagglutinin (HA) that sits on the surface of flu viruses. The head mutates easily and frequently and therefore those flu vaccines targeting the head must be continually re-formulated to ensure they are effective. In practice, the flu virus, therefore, is often one step ahead of the production of vaccines which therefore often are seen as ineffective.

In two studies, separate research teams described how they created novel flu vaccines that target the 'stem' of the HA molecule instead of the head.[32] The stem of the HA molecule is similar across different flu strains and mutates far less often. One of the teams, led by Barney Graham at the National Institute of Health in Maryland, created their vaccine by attaching part of a flu virus's HA stem to tiny balls of protein. These protein nanoparticles kept the stem intact and made it easy for the immune system to spot once it was injected. A second team, led by Antonietta Impagliazzo at the Crucell Vaccine Institute in Leiden, created their own experimental flu vaccine by removing the head of the HA molecule, and tweaking the stem to make it bind to antibodies more effectively.

The *Guardian* newspaper website reported, "Influenza remains one of the most serious public health challenges, and new therapeutic and preventative solutions are needed," a quote from Hanneke Schuitemaker, head of viral vaccines discovery at Janssen Infectious Diseases and Vaccines, a company that worked on the vaccine. *The Guardian* also reported[33] that Sarah Gilbert, professor of vaccinology at Oxford University said, "this is an exciting development, but the new vaccines now need to be tested in clinical trials to see how well they work in humans. This will be the next stage of research, which will take several years". So we are still some way from having better flu vaccines for humans.

Antibiotics. The problem in developing new antibiotics is that they are not very profitable, compared with, say, a drug to control high blood pressure (*hypertension*). This is so because an antibiotic, if it works, puts itself out of a job, i.e. it is a short term fix. The hypertension drug, in the other hand, tends to be for the rest of the life of the patient. The result is that all the major drug companies have recently (2018) cut back their teams working on the development of new antibiotics.[34] One of the consequences of that cutback may well be the loss of workers experienced in developing new antibiotics; clearly a tipping point sort of danger. Despite these difficulties, there are signs that government and, in particular, smaller drug companies, are picking up the challenge and there are now nearly a hundred new drugs in trials. Some of these drugs are tackling routes to kill bacteria in new ways. Some will come through. So, there are good signs and we must press on with them.

The important point being made here is that we really can do something about these under-lying problems of population and the need for food and energy. However, there are complications and global population growth is fundamental to the current generation of farmers' function of supplying global food and energy needs. So, we can do at least some of these things but that leaves the questions of will we actually do those things and will we do them in time?

If we do not do these things, then it may be that population will peak and then collapse to maybe only a couple of billion before it stabilises at whatever figure, maybe three or four billion, maybe one billion or whatever figure, but it won't be twenty billion or

32. Impagliazzo et al: *A Stable Trimeric Influenza Hemagglutinin Stem as a Broadly Protective Immunogen*, *PubMed*, August 24th 2015

33. Sample, I: *Universal Flu Vaccine a Step Closer as Scientists Create Experimental Jabs*, *The Guardian*, August 24th 2015

34. McKenzie, D: *Facing the Resistance*, *New Scientist*, January 19th 2019, pp20–21

whatever the peak might otherwise turn out to be.[35,36] Populations do not stabilise at their peak; they _always_ collapse back to a lower level. The logic confirmed by population studies indicates quite conclusively that stable population levels are a fraction of the peak. When the collapse happens, and it appears logical that it will, then it will be painful. It will be painful if you are one of the ones that die. It will be painful if you are one of the ones who survive because the world will be a very different place after that fall. Many of the things we take for granted will not be there anymore; the fact that the human race used to have the knowledge of how to make a garden spade does not mean you can have one. Could you go and make steel on your own?

Now, if you are a politician, it is perhaps not a good idea to be in charge when the collapse becomes obvious as imminent. Better to be prepared to try and be one of the nations that planned to avoid it. Secondly, there is something perverse about any catastrophic fall in global population; we will dramatically cut greenhouse gas production. So, it may be that catastrophic collapse of population due to disease will cause a collapse of numbers in the human race before global warming makes the world too difficult for current population levels. Either way, there will be a fall. Until that point, what has this to do with global warming? The answer is about energy use.

Food Production – Hunger and Obesity

It is not just the extremes, but the billions in between we have to cater for. Over-feeding, (obesity) is an expensive killer. The killing may help the population numbers issue but it is far from the best way to cut numbers. At the other extreme, starvation is not a nice way to die and it tends to cause migration and war.

Obesity

In the developed countries, particularly the West, obesity is a health time bomb with potentially catastrophic effects. It is well recognised that overweight and obesity increase the risk of this country's biggest killer diseases – coronary heart disease and cancer – as well as diabetes, high blood pressure and osteoarthritis.[37]

Even as long ago as 2009, the National Audit Office (NAO) found that obesity is responsible for more than nine thousand premature deaths each year in England and reduces life expectancy on average by nine years. Obesity also has significant financial costs, both to the NHS and the wider economy. In common with other countries around the world, levels of obesity in England are rising. The consequences are serious.

What obesity does, in the contest of the triad being discussed in this text, is add to the pressure on health, food and population disease risk, i.e. pandemics. Perversely, obesity decreases life expectance and, therefore, helps reduce population growth. As ever, these situations are hyper-complex.

35. Stanton, W: _The Rapid Growth of Human Populations 1750-2000_, _Multi-Science Publishing Company Ltd_, 2003

36. Butterworth, B: _Reversing Global Warming for Profit_, _MX Publishing_, London, 2009

37. _Obesity: Defusing a Health Time Bomb_, _National Audit Office_, May 1 2009

Hunger

Population growth is out-stripping food supply. It has been so for many centuries but population growth is now exponential and food supply is not. The Feb 2015 issue of *The Furrow* from John Deere showed, on page 15, a stunning histogram of World Bank/United Nations prediction of population growth and available farm land up to 2050. It is reproduced below and it is a grave warning of the human race sleep-walking to disaster. If you are under sixty years of age, you have a pretty good chance of living till 2050.

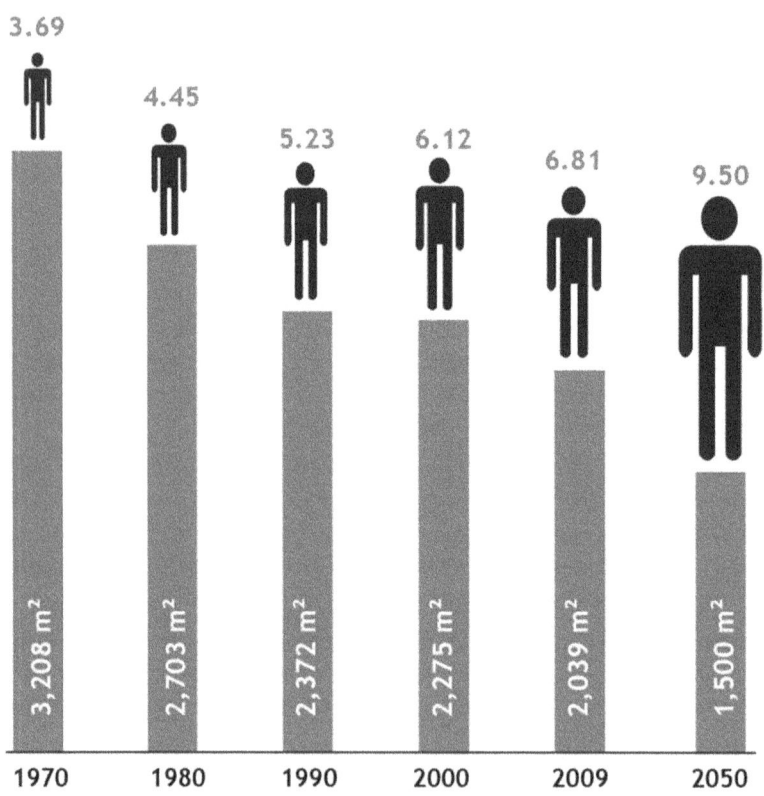

FARMED AREA PER CAPITA COMPARED TO WORLD POPULATION

WORLD POPULATION IN BILLIONS

Sources: The World Bank, World Development Indicators 2010/2011

Food and Agriculture Organisation of the United Nations

Achieving sustainable gains in agriculture

Fig 4.7: Predictions on the Amount of Farm Land Related to the Size of the Global Population

The energy needed to manufacture and deliver mineral fertilisers, especially nitrogen,[38] is a major share of global energy consumption and a real dilemma in planning. It is also a potential vulnerable spot for individual farms in terms of cost and pollution of groundwater. However, we most certainly do need mineral fertilisers, and more of them, especially nitrogen if we are going to seriously try to feed the global population. However, we also need to recycle wastes to land and this will raise organic matter in soils. In turn, raising soil organic carbon reduces losses of mineral nitrogen to groundwater. Indeed, if mineral fertilisers are blended into the late stages of a composting operation, then this will reduce, maybe eliminate, the potential losses of nitrogen to groundwater by rainfall or irrigation at a later date.

Bear in mind that one tonne of nitrogen nutrient in mineral fertiliser, made in a modern USA factory does, according again to United Nations sponsored research[39] take 21,000 (yes, no mistake, twenty one thousand) kWh of electricity to manufacture and deliver. Some factories in production elsewhere in the world are up to a factor of twenty times less efficient. Global nitrogen fertiliser production is approaching 150 million tonnes. Most of that electricity will have come from burning fossilised, hydrocarbon fuels (see www.landresearchonline.com December 20 2014). As a guestimate, that may be *more than* a global consumption of a thousand, trillion kWh electricity per annum (that is 1000,000,000,000 kWh).

Putting it gently, '*Houston, we have a problem!*' We need to:

- *Stop* population growth (just reducing growth is not good enough).
- *Learn* to be vegans and preferably vegetarians. Meat production takes a lot more land.
- *Stop* building on good agricultural land.
- *Stop* mono-cropping energy crops.
- *Find* new sources of fertiliser which are not energy based (see *Section 4*).

If we are going to have sustainability, then we have to reduce energy inputs and increase output at one and the same time.

In developed countries there is an unhealthy habit of eating far too much meat and milk products. It reduces well-being, increases health problems and there-by increases health costs to the state, and shortens life. As Le Page reports[40] in the *New Scientist*, "...it is hard to see how diets could be radically changed without imposing a carbon tax". This is one of those issues which politicians, fearing a back-lash from their constituencies, fear and avoid. That is not how we are going to crack this problem.

There is one more complicating factor in expanding population and the need for food production to match. It might well seem logical that, as carbon dioxide content of the atmosphere rises, there will be at least a little extra growth from crops. However, there is some research which suggests that while such increased production may occur, the crops involved may have a lower content of vitamins and minerals.

38. Gellings, CW & Parmenter, EC: *Energy Efficiency in Fertiliser Production and Use, Efficient Use and Conservation of Energy Volume 1, UNESCO*, 2004

39. Ibid.

40. Le Page, M: *We Have to Pay the Real Price of Food, New Scientist*, January 26th 2019, p25

Energy consumption

Farming is a major consumer of energy on the global scale and in any developed agriculture, farming will take a very significant proportion of the nation's total energy consumption. Agriculture requires energy as an important input to production. Agriculture uses energy directly as fuel or electricity to operate machinery and equipment, to heat or cool buildings, and for lighting on the farm, and indirectly in the fertilizers and chemicals produced off the farm. In 2002,[41,42] the US agricultural sector used an estimated 1.7 quadrillion Btu of energy from both direct (1.1 quadrillion Btu) and indirect (0.6 quadrillion Btu) sources (a total of around 500 GWh). However, agriculture's total use of energy is low relative to other US producing sectors. In 2002, agriculture's share of total US direct energy consumption was about 1%.

Do these figures mean anything to anyone other than governments and statisticians? Probably not but they are significant because they are very large and indicate not only a cost to farm businesses but also a hidden cost to environmental stability in the resultant production of greenhouse gases and global warming with, in turn, having an effect on closed loop farming.

Waste Production

Waste is the potential mirror image problem that we need.

There is a relationship between human life and the wastes produced. It is sometimes said that it is possible to judge the wealth of a country be weighing its garbage. It sounds like a glib statement but there is some truth in it. There are, of course, varying estimates of the production of municipal waste per capita and, of course, these figures are rising. At the time of writing, the OECD (Organisation for Economic Co-operation and Development) figures indicate that one person's output of municipal waste per annum is:

Country	kg per person, per annum
USA	760
UK	560
Japan	410
China	120
India	100

These figures give some idea as to the opportunity to use wastes for land (including deserts) reclamation, for making fertilisers and growing crops. However, the available statistics generally are limited to *municipal solid waste* (MSW) and give neither the potential soil-useful content of that waste, not include many other wastes from industry which might be of agricultural value. Globally, there are billions of tonnes of industrial

---

41. Outlaw, JL et al: *Agriculture as a Producer and Consumer of Energy, Agricultural and Food Policy Center, Texas A&M University*

42. Schnepf, R: *Energy Use in Agriculture: Background and Issues, CRS Report for Congress,* November 19th 2004

wastes which contain organic carbon in a form which aerobic micro-organisms can process. These include, just as a few examples, liquid plastics such as PVA (polyvinyl alcohol), paper and cardboard wastes, wool-based carpets, filter-trap wastes, wastes from food and alcohol production, animal body tissues, sewage and so on. See *Appendix* for a longer and more specific list.

What is clear is that there are billions of tonnes of waste around the world which could be of value in playing a role in re-balancing at least part of the triad.

Hospital Wastes

In the UK, and apparently in most countries in the world, hospital toilet wastes go directly into the public sewer. It is relevant to note that this is, quite legal (but questionably safe) in the UK at least. The dilute liquid and solids are put directly into the public sewer system and treated, along with general urban sewage, in conventional public sewage treatment works (STWs) with the products released to water courses (some of which are extracted downstream for human consumption) and to fertilise farm fields. During transit to the sewage treatment works, the dilute liquid does, of course, contain significant populations of disease organisms and dilute anti-biotics and drug residues. This environment is likely to kill off some of the pathogens but, in so doing, selects out the more resistant ones. What is not known is to what extent this mechanism is a significant one in actively developing resistant strains of bacterial pathogens. It seems likely that this may be the case. There is a further indicator of risk. It has been known for over thirty years that residues of the hormone used in the contraceptive pill have been tracked and are urinated out and, again, enter the public sewage system, go through the system and end up in the rivers where small, *male* fish have produced eggs. Indeed, this has continued to build up so much so that Professor Charles Tyler,[43] University of Exeter, is of the view that over 20% of river fish are now transgender. His findings were presented at the 50th anniversary symposium of the Fisheries Society in the British Isles, held in Exeter University 3–7 July 2017. It is known that clinical residues do get though the conventional sewage system and recent research confirms a link to cancer in wales. In a recent issue of *New Scientist*[44,45] there was a report of some Australian research which identified sixty-nine *named* drugs which had got through the sewage collection and treatment systems into river bed muds and back up the food chain. Here in the UK, our population density is higher than Australia and the risks, therefore, likely to be also higher and more complex. Sewage from a hospital is likely to be the most concentrated of all waste sources of drugs, hormones and pathogen residues. The extent of risk is only superficially known but may be significant. Also note the increase in numbers in what is undoubtedly a complex situation of apparently trans-gender/mixed up children and wonder if there is a connection.

---

43.   Tyler, C: *A Fifth of Male Fish in UK Rivers Now 'Trans-Gender' Due to Chemicals in Human Waste*, <u>University of Exeter News Letter</u>, July 3rd 2017

44.   Ibid.

45.   Klein, A: *More Than Sixty Prescription Drugs Are Getting Into River Foodchains*, <u>New Scientist</u>, Oct 6th 2018

Vaccine and Antibiotic Failures

Whether western nations like to admit it or not, their health services are mainly driven by the pharmaceutical industry and their need for profit. It is important to add that this has produced some truly staggering results. Further, running parallel to that is a widely held faith that anti-biotics will cure just about everything and will always be there. They will not do so for much longer.

Despite these difficulties, there is, as ever, hope in the ingenuity of the human race. There is a significant amount of work being done on an alternative and that is to help the human body develop its own antibodies and resistance to disease. According to the *New Scientist*, the CRISPR genome editing revolution continues to advance at an astounding pace. As many as twenty human trials will be under way soon, mostly in China. One of these trials will involve the first-ever attempt to use CRISPR to edit cells while they are inside the body. The aim is to prevent cervical cancers by targeting and destroying the genes of the human papillomavirus (HPV) that cause tumour growth. This study is due to begin in July at the First Affiliated Hospital of Sun Yat-Sen University in China.

Gene therapy, which involves adding extra genes to cells, was first used to cure people in 1990, but it is mainly useful for treating rare genetic disorders. In contrast, gene-editing, which involves altering existing genes inside cells, promises to treat or cure a much wider range of conditions, from HIV infection to high blood cholesterol.

Chapter 5: Conclusions So Far

The Triad: Global Warming, Population Growth & Antibiotic Failure

There is no doubt that the triad of global warming, population growth and potential vaccines/anti-biotics failure is the biggest potential disaster that the human race has ever faced.

Firstly, there are too many people. We could deliberately kill off, say, five or six billion and that would take the pressure off. There might be some difficulty in choosing the five or six billion to go and they just might not go quietly. Alternatively, all people over, say, sixty-five years old could be put to sleep. There is a logic in saying that if we do not do it, nature will do it for us.

Secondly, increasing wealth produces more waste and pollution. So, if the five or six billion people above were selected as the most wealthy, that would help. They might object and have the power to do so.

Thirdly, if anti-biotics do not keep up with disease levels, nature will do it for us but probably of the five or six billion poorest.

There is here a question of ethics and there is some guidance in the Dalai Lama's *message to the world*.[46] His Holiness is an incredibly practical man and argues that there is something that transcends all religions, i.e. ethics. Taking that as a starting point, individuals, politicians, scientists, all of us, might raise a question this way; Is it more unethical to suggest deliberately killing off five or six billion people (young or old, rich or poor) or to not face up to the facts as best we can see them at this point in time and at least try to do something about it before nature does it for us – which will be even more painful for more than five or six billion people?

Is this too abrupt an analysis? Maybe, maybe not. Generally, the human race is not good at facing up to difficult situations, especially *en mass*, demanding co-operation and rapid action. However, the clock is ticking.

The Complication of Aging Populations

The 'one-baby-per-family' policy of China has done part of what was necessary for the Chinese economy to catch up with population growth. India please note. However, the authorities in China are now having problems with an aging population which relies on young people to care for the aged. The aged are becoming a growingly obvious burden on the younger, working population. So, the authorities want couples to have more children. What seems to escape the powers that be is that, not far into the future (i.e. every day from now on), the same problem will occur, except it will be bigger. There is an answer to this problem, taking on the theme of the above paragraph; take the aging population, say above seventy years old, and put them to sleep quietly. The reason why

---

46.   Dalai Lama & Alt, F: *An Appeal to the World*, *Harper Collins*, 2017

all of this is put so bluntly is that there really is a problem and just kicking it down the road just makes it worse. We have to find ways to limit population more urgently than it is already happening, and we have to find ways of managing the imbalance between those that are working and those that are not.

It is not logical to solve the aging population problem by increasing the population.

What Action is Needed?

Clearly, if the argument about the convergence of the triad is accepted, then global warming, population and the potential failure of anti-biotics have to be tackled. There is urgent need to;

i) Increase the rate of change to renewable energy, particularly where there is a good life-time return on the original energy investment, and also in new technologies such as wave and tidal power.

ii) Limit population growth and put it into slow reverse by 2030. Face the problem of aging populations by tackling age and social structure related to work, i.e. not by encouraging birth rate to climb.

iii) Develop attitudes to healthy living, extend the use of existing antibiotics by restricting their use to only genuine, proven therapeutic use, and develop new antibiotics.

iv) Develop and enable zero waste society by avoiding waste production, re-use at the point of production and recycling. Enable local small scale recycling on a global scale (see *Chapter 7*).

By whom
Everyone. Every individual and every part of every society globally.

You and other individuals
A good place to start. Actively influence government at all levels. Be part of it all.

Local communities
Happiness research is complicated because the situations researched refer to individuals who tend to be subjective. Nevertheless, Dan Beutner[47] looked at the people of the town of San Luis Obispo in the USA and saw that a community could organise themselves in terms of opens spaces, shopping hours, participation in municipal affairs, how people talk to each other and become happier. So communities can do things themselves and influence others. It is not impossible for communities to tackle the triad discussed here.

Cities
Cities have the power to change the lives of millions. The local planning authority can dramatically influence building design and energy use. Cities organise the collection of wastes and what happens to it.

---

47. Beutner, D: *Thrive: Finding Happiness the Blue Zones Way*, *National Geographic*, Washington, 2010

Multinationals

Large companies span boundaries and can influence governments. They can dramatically affect the issues involved in the trial of risks discussed here. The oil companies need to be involved here and be seen to genuinely assisting survival.

Governments

Not only can governments address these issues, not only do they have a responsibility to tackle them, the responsibility of the safety of their citizens is their highest responsibility.

UN

The UN has an enviable record of achievement in very difficult circumstances. There is more to be done in influencing the powerful countries, and multi-national companies, to do more on every count but especially this triad of pressures relating to the nominal date of 2030.

Section 3

The Options

Chapter 6: Why Do Anything?

The leader in the *New Scientist*, October 13 2018 said, "the natural reaction to the IPCC report and wider developments may thus be despair. But that guarantees only one outcome: defeat. As the report makes clear, we still have time to pull off a rescue. It will arguably be the largest project that humanity has ever undertaken – comparable to the two world wars, the Apollo programme, the cold war, the abolition of slavery, the Manhattan project, the building of the railways and the rollout of sanitation and electrification, all in one. In other words, it will require us to strain every muscle of human ingenuity in the hope of a better future, if not for ourselves then at least for our descendants. The history of humanity facing a crisis is one of blind stupidity, denial and dawdling followed by heroic rear-guard action to prevail against all odds. The climate crisis is close to that inflection point. Does our generation have the gumption? It is time to find out."

In terms of action, putting it bluntly, there are three choices here;

i) Reverse population growth. If we do not do it first, and urgently, nature will do it for us.

ii) Reduce the global dependency on fossilised fuels, urgently and dramatically.

iii) Replace mineral fertilisers with digested 'wastes'.

The first two of these choices will be opposed by very powerful vested ('natural' or financial) interests. The third will involve a change in attitude to risk (from emotional to science-based) by governments and regulators.

Chapter 7: Electric Cars and Nuclear Power

Electric Cars

The negatives on cars first. We really do need to be careful about electric cars and *hybrids* (using a petrol or diesel engine coupled to an electric motor and batteries).

A car run on only an electric motor may be, in itself, 'clean' and good from the point of view of local pollution. However, where the electricity comes from, often from a remote power station many miles away, generating its fuel from burning something, probably fossilised fuel and probably with a very poor total system efficiency, is hardly a sustainable solution. Frankly, a hybrid is probably a better solution. In either case, the life-cycle energy cost of the batteries including what to do to dispose of them, or recycle them, at the end of their useful life, is another consideration which is often over-looked and likely to be a negative. Having said that, the available information is that producing the electricity in a central power station from fossilised fuel is slightly more efficient than burning it in each car separately.[48] While on the face of this, it looks as if this helps break our dependency on fossilised fuels, it has to be remembered that most of the energy used in the life-cycle of a car is in its original manufacture, not in the fuel consumed while driving it about in our daily lives. So, the solution is to wean us off cars and life near work and shops.

Nuclear Power

Until recently, most thinking and informed scientists had serious reservations about nuclear power. There is some move towards nuclear power on the grounds of reduced carbon dioxide emission and some scientists see nuclear power as less of a risk than global warming. That is about nuclear power from fission. What about fusion?

Over sixty years ago, the BBC television news announced that British scientists has succeeded in harnessing the power of the H-bomb with a process that had been christened *ZETA* – Zero Energy Thermonuclear Assembly. The plus is that this is 'clean' energy. They are still working on it. The problem is that the centre of the nuclear reaction is so hot, no material will contain it and, so far, containment is by magnetic field. However. There is a new example of global co-operation to progress the technology. In southern France, thirty-five nations are collaborating to build the world's largest tokamak, a magnetic fusion device that has been designed to prove the feasibility of fusion as a large-scale and carbon-free source of energy based on the same principle that powers our Sun and stars. The experimental campaign that will be carried out at ITER is crucial to advancing fusion science and preparing the way for the fusion power plants of the future.[49] The ITER website claims:

---

48. *Energy consumption, CO_2 emissions and other considerations related to Battery Electric Vehicles, European Association for Battery Electric Vehicles*, April 8th 2009

49. *ITER website*

> ITER will be the first fusion device to produce net energy. ITER will be the first fusion device to maintain fusion for long periods of time. And ITER will be the first fusion device to test the integrated technologies, materials, and physics regimes necessary for the commercial production of fusion-based electricity.

Well, we all hope so. The project was launched in in 1985. The ITER members – China, the European Union, India, Japan, Korea, Russia and the United States, are now engaged in a thirty-five year collaboration to build and operate the experimental device.

That project, however is not the only one to aim at harnessing fusion power. Tokamak Energy is another project.[50] This, if and when it works, is relatively small device and Tokamak is thinking of groups of machines on 'Tokamak farms'. Others have other ideas. Says physicist, Harrold M Butterworth, "These machines are intrinsically safe and the only very long term problem might be the reaction vessel's liner that will have seen prolonged and massive bombardment with neutrons. Nevertheless, once proved there will be a scramble to build them. If at least some of this stuff works the CO2 problem is greatly reduced because there will be lots of energy available for everyone without any need to burn anything. Ever since the first attempts back in the 1940s to extraordinary promise of power offered by $E=MC^2$, fusion energy has been 'within thirty years'. At last, that is changing. It may be 2035 or 2040 before it is in very wide use, but there are very real signs that it is becoming reality."

It may sound unkind, but we need, however, to do something dramatic in less than thirty-five years; *much* less. The solution is a long way off commercialisation and not a contender in terms of the timeframe we face on global warming.

---

50. *Tokamak Energy*

Chapter 8: Transition Fuels

Unconventional Gas – Shale Gas

Shale gas is a hydrocarbon fuel, so it does produce carbon dioxide when burned, so it would be better not to use it if we have a sustainable alternative. We do have alternatives (see below in this section) but not enough, fast enough. The other hydrocarbon fuels not only produce carbon dioxide when used, but also other pollutants, including nitrous oxide, which is a much more potent as a GHG than carbon dioxide. The worst polluter in wide use, especially in China and also in Germany, is lignite, or brown coal. Less damaging are the hydrocarbon oil fuels including petrol and diesel. Petrol is cleaner than diesel when burned in a car engine but burns roughly twice as much per mile/km as diesel, so produces twice as much carbon dioxide. Where shale gas fits in is that is almost a clean burn, producing carbon dioxide but not much else. So, provided shale gas is used instead of the other hydrocarbon fuels, it does have a value. As a 'transition fuel' it has a place but we really do need to shift away from all hydrocarbon fuels as quickly as possible.

The global exploitation of shale gas exploration and production will be both at sea and on land. Where it is on land, it will affect the landowner where the drilling takes place. Different countries have different laws about the ownership if mineral rights. For example, in the USA it is normally the case that the land owner owns the mineral rights (including oil and gas) below that land. In the UK, alternatively, that is not the case; the state owns the mineral rights and licences them to businesses to exploit them with or without the approval of the landowner above.

The Basics

The vertical shaft of a drill exploring for shale gas in the USA might have been less than 500 meters (around 1600 feet) and through relatively soft rock. In the UK, the 'top hole' is quite likely to be a kilometre, maybe two, or (in old money) a mile or more and through harder rock. Clearly, global situations vary. That, in itself, is not that much of a new thing. Deep drilling for all sorts of reasons (such as geothermal drilling to bring 'free' hot water to heat homes, offices and shops) has been going on for a long time. What is different about drilling for shale gas is that when the vertical shaft has got to the depth that the geologist thinks is right, the drill turns in a long 'J' formation from being vertical to horizontal. In the horizontal bit, the engineers want the hole to leak – *inwards to collect the gas*! To assist that, *hydraulic fracturing* (with high pressure fluids) is used. The top hole will be a sealed telescope lined with concrete and steel down to many hundreds of meters.

Risks and pollution management

There is no such thing as 'no risk' anywhere in this universe. For example, water is essential to most if not all forms of life as we know it but heavy rainfall can cause flooding and too much water for a human will cause drowning. That principle of there *always* being a risk applies to the exploration and production of shale gas as much as anywhere or anything else. The question about risk is not if it can be avoided, it is about whether we have adequate knowledge of what the risks are and if we can manage them. There is another risk which is often forgotten by those who object to any particular event of proposal; what happens if we *not* do this. With apologies to Shakespeare, "To drill or not to drill, that is the question" – but see the whole package, not just a bit of it.

Earthquakes

A special report on Sky News UK, April 19th 2015, looked at a relationship between fracking and earthquakes in Texas. That report portrayed Professor Brian Stump, SMU University, Dallas, quietly saying that "there is undoubtedly a correlation between fracking and earthquakes". The professor is an acknowledged expert seismologist. Not taking his statement seriously would be foolish. However, what does it mean?

Many years ago, when I was a student learning about statistics, it was pointed out that there is a *correlation* between the divorce rate in the UK and the rate of importation of wooden rolling pins. The point is that it is not difficult to find sets of figures which have similar patterns and therefore appear to be related. They might be. They might not be. What the word 'correlation' does is to alert that there *might* be a relationship. To establish that there is a real connection, investigation must show that there is a logic or understood mechanism involved, and there has to be evidence. When I used to do a great deal of work as an expert witness in court, if I wanted the judge to believe me, then I had to state a clear opinion, give a common sense (logical) explanation for that view and show evidence that is was true. The essence of scientific thinking is to have a hypothesis, have a logical explanation, and be able to test to provide evidence. Frankly, Professor Stump is dramatically more competent than I to comment or test that correlation. Nevertheless, I am left wondering and it may be that this is so complex, or at least so variable from one location to another that we will never be able to be definitive. It seems logical to observer that there are, from time to time, new areas of seismic activity where none had been recorded before or we now have more sensitive recording equipment. These may be that nobody had bothered recording before (so there was no record) or they had recorded no activity but, all of a sudden, there was some – that may be due to the earth itself which is still very hot at its core and still cooling. So, if there was a situation where in the USA, as Professor Stump observed, adequate records showing more seismic activity after shale fracking than before, it *may* have been due to that fracking, it *may* not. Until more is known, it seem to me that the following is a reasonable position to take. Nevertheless, we would do well to note the observation and be careful and forever recording and trying to understand fully.

The first, widely publicised shale gas well drilled in the UK was near the Northern seaside resort of Blackpool. There were reports of earthquakes and the drilling of the shale well was immediately blamed. You don't have to be a qualified seismologist to understand something about fracking and earthquakes. Consider a few billion tonnes of rock a mile or two below the surface and put some water under pressure into a crack in that rock. What does common sense tell you is going to happen?

For those old enough to remember the Cold War between the West and the Soviet Union, each testing bigger and bigger hydrogen bombs usually below the ground, they may remember that there was a fear that we might crack the surface of the earth's crust. These weapons were hundreds, even thousands of times larger than the bombs dropped on Japan during the Second World War. Hiroshima's bomb was around fifteen kilotonnes (equivalent to 15,000 tonnes of TNT explosive) and the USSR's largest 50,000 kilotonnes. Hydraulic fracturing is just not in the same league.

Blackpool has always had earthquakes. These have been noticeable maybe two or three times a year but never done any sudden damage. Millions, even trillions, of tonnes of earth's crust is under tension and stress and, occasionally, it moves a bit. Now, if that tension and stress was on the edge of moving and causing a detectable quake, what fracking might do is trigger the quake which would have happened anyway. More than that, what the injection of fluids are more likely to do is lubricate the movement which was about to happen anyway and make a series of short movements which would reduce the effect of the quake. Blackpool might well have less detectable earthquakes, not more.

Sensibly, we do need Professor Stump and others to keep an eye on this and keep investigating. Nevertheless, at present, the evidence we have is that this, *in the UK*, is at worst not a serious risk.

Pollution Management
The most common objection to shale gas exploration is pollution risk. That comes down to a simple analysis; the risk depends on what gets put down the hole, what is drilled through and how what comes out of the hole is managed. Add to that whether the vertical drill-way, or 'top hole' can be safe.

One of the risks in drilling is what might be drilled through and whether that either make drilling difficult or actually dangerous. There is a risk in drilling (for any purpose) in some geographical areas where radon gas, which is radio-active, could be released from strata being drilled through. The gas is released to the surface, quite naturally and without any form of drilling, in some parts of the UK for example, including Cornwall. There, it may collect in unventilated cellars and, if and when it does, it is potentially carcinogenic. Frankly, in drilling, such possibilities will be considered at the stage of providing the risk assessment for (in the UK or whatever authority there is in any country in the world) the Environment Agency and Health and Safety Executive and there is understood, low and managed risk to human health, wild life or the environment.

Common sense tells us that whatever the pollution risks are of leakage from a mile or so down back to the surface, they are very, very small. In practice, it just is not going to happen for one very simple reason. If it was going to happen, it would have done so already during the last few hundred, million years.

That still leaves the worry about the integrity of the vertical shaft. This shaft is like a telescope of concrete with steel pipes inside. Each stage of the telescope can be pressure tested before progressing further to build the next, lower, stage. This vertical shaft certainly might travel through strata near the surface which might leak back up to the top. It certainly might be drilled through aquifers which might be used for human consumption; leakage certainly might cause pollution. How likely is that 'might' and can it be controlled?

Leakage of the vertical shaft after construction is known but it is rare. After all, sinking just the vertical shaft is quite likely to cost over US$15 (GBP £10) million in the UK and, therefore, the investors and engineers are going to be quite careful. The way of covering this risk is to pressure test the vertical shaft before turning to the horizontal drilling. If it leaks, abandon it.

That, however, still leaves the drilling and construction of the vertical shaft through a range of strata, some of which may be porous. It might well go through an aquifer. How is this risk covered? Well, in the USA it may not always have been considered. There has, undoubtedly, been pollution from unscrupulous operators there. Regulation and inspection could, in theory, identify, control and eliminate this risk. In the UK, this would almost certainly happen. That leaves the question of exactly how to control the risk itself. Of all the countries in the world, in the entire history of the human race, the UK is the most regulated *ever*. That still begs the question of what is the fail-safe technology which the regulators must look for?

The answer is quite simple; use only a drilling fluid which is completely environmentally friendly; then it does not matter if the vertical shaft leaks. Do such fluids exist? Well, at least one is licensed under the water supply regulations for drilling through aquifers for human consumption which, quite literally, means you could drink it.

Drilling Fluids – Also Known as Drilling Muds
Drilling fluids, often called 'drilling muds' in the industry, have a number of functions. They are usually pumped down the shaft that operates the drill bit itself in order to lubricate and cool it. That is the first function of the fluid. The second is to flush the cuttings away from the drill bit and back up to the surface to be recycled to good use elsewhere.

Those functions are obvious enough but there are others. The pressures in the fluids are often high, for example, imaging a drill hole of maybe 2000 metres. Now, ten metres of water is approximately one atmosphere pressure, so 2000 metres is 200 atmospheres. At these pressures, the wall of the hole may distort and possibly fracture and leak. Therefore, additives are needed to stabilise and seal the walls.

At the depths of 2000 metres or more, temperatures may be very much higher than at the surface and bacteria and fungi may develop. In such cases, it may be necessary to add an appropriate sterilant and this may affect what happens to the flow back when it gets to the surface. However, the concentration in the spent fluids which may be discarded or re-cycled is likely to be very low and will be metabolised easily in the agricultural soil, forestry soil or bio-processing such as composting.

Laterals
The technology of the laterals has advanced very quickly. At the time of writing, it is possible to drill maybe a dozen laterals from one vertical well, each lateral up to ten kilometres long (six miles) and no doubt, fifteen kilometres (ten miles) will happen and more. It is the area of these laterals that the fracking takes place and it here that there is concern, maybe justified.

While there is evidence that there is safe and tested technology for drilling of the top hole for shale gas, it has to be admitted that the technology of managing the lateral drill-ways and production through them is more complex and is at the frontier of environmental management. The most commonly voiced risk here is in the production of millions of gallons of brine and how that is managed. Well, the county of Cheshire sits on top of enormous quantities of salt which has been mined for centuries. Yet, the groundwater and the people on top are OK about it. In terms of shale gas exploration, there is new technology about; it is called electrodialysis. The process itself is not new but it has previously been economic only for very low concentrations of sodium chloride (*brine* when in solution in water). However, researchers in MIT (Massachusetts Institute of Technology in the USA) working with the KSA (Kingdom of Saudi Arabia) have shown that a multi-stage process can be economic and allow the recycling of the water – so dramatically reducing water demand and the risks involved in what to do with the spent liquids. There is another way of dealing with brine and that is to put it on sugar beet crops. Farmers buy what they call *muriate of potash* or just 'potash' to put on most crops. That potash is, in fact, potassium chloride and most crops would very much respond to that and will respond also to sodium chloride (common salt or brine) but, with most crops, the sodium salt becomes toxic and inhibits growth at quite low application levels. However, beet crops can handle the sodium salt and grow just as well on it – so famers put common salt to fertilise beet. There might be some liaison here with shale gas drilling.

There is some really exciting stuff coming out of the research centres round the world, much of it from the USA, on microbial activity generating electricity while processing wastes. Much of this research is to do with processing organic wastes – containing carbon from which the micro-organisms get their energy. However, there is an exciting application possibility with shale gas production. In the drilling and exploration for shale gas, large quantities of brine are used. When this comes back to the surface, the liquid which is mainly water, also carries significant amounts of hydrocarbons (as oil and dissolved gas). Those hydrocarbons can be used by the micro-organisms to produce electricity to power a battery made of the brine which is, of course, ionised. So, the sodium goes to the cathode and the chlorine goes to the anode. Furthermore, say the University of Colorado Boulder, there is a surplus of electricity after running the desalination process.[51]

This is certainly break-through technology but, one must suspect, there is a long way to go before this can be scaled up and widely used. Nevertheless, in principle, there is reason to believe that we do now have the technology to manage the drilling process and production operations for shale gas in a way that is environmentally sensitive.

Spreading Cuttings and Spent Fluids to Proximity Land
When asking "what comes out of the well?" the answer is, hopefully, mostly gas. Before that happens, and indeed for the life of the well, lots of other things come out. Much of the other stuff is potentially not environmentally unfriendly.

Firstly, the shaft will be bored through a range of strata, some of which may contain elements or compounds which might be toxic in some way. The drilling fluids which are used to carry the drill cuttings out of the well as it is bored (the *flowback*), will also bring out these other materials – if they are there. Secondly, the high pressure water

---

51. Stoll, Z et al: *Shale Gas Produced Water Treatment Using Innovative Microbial Capacitive Desalination Cell, Journal of Hazardous Materials* 283C:847-855, October 2014

used in volume to create the hydraulic fracturing will also dissolve materials from the shale, especially sodium chloride – common salt. Anyone who dismisses these potential dangers is, at best, irresponsible, and at worst, criminal.

There are two possible approaches to dealing with these 'arisings' out of the well. It is important to note that in the UK (and indeed all of the EU) these arisings are legally a *controlled waste* and that means subject to regulation – of which there is plenty and the Environment Agency knows that they will be watched every step of the way by a lot of aggressive people (some of whom are emotional, not very well informed and motivated by overseas interests).

The first way of dealing with the arisings is to isolate them in a restricted area. That could be in a lagoon or enclosed space and left there forever. Alternatively, the cuttings could be used in, say, the construction of sea wall and flood defence work. It is likely that our regulators will favour this route because it is relatively easy for those drafting the regulations to identify the risks and write the regulations to contain the risks – even if it means permanently. The disadvantage of this route, hover, is that if there is a concentration of a material which might be toxic, it is still there as a 'point risk'.

The second way is to remember that nature is remarkably resilient and, given time and enough spreading out, will deal with almost anything and to its sustainable advantage. This route might be favoured by environmental scientists with the right training and experience because it provides for the identification, management and the sustainable elimination of the risks by creating an environmental benefit. The advantage of this route is that if (again *if*) there is a concentration of a material which might be toxic, then a *dispersed risk* can be identified and managed by competent people and processed out of danger.

There are very good reasons why the arisings of cuttings and spent fluids should be put to proximity land. Firstly, farming in the UK loses somewhere between two and three million tonnes of topsoil to the sea by erosion every year. For those fond of a 'tipple' of prosecco, in the traditional centre of prosecco production, between the towns of Valdobbiadene and Conegliano, a new analysis estimates that 400,000 tonnes of soil is lost by erosion every year in the vineyards. The region produces ninety million bottles of prosecco annually, meaning that 4.4 kilograms of soil is lost for every bottle of fizz.[52] The point here is that we are losing topsoil to the sea at an unsustainable rate, continuously. Where we can use drill cuttings to replace some of that, it is logically a plus.

The cuttings are of diameter of sand down to nothing. It would be sensible to use them to replace some of that lost top soil. Secondly, as box one below indicates, this would get a very large number of truck movements off the public roads. Thirdly, these materials which are legally branded as 'waste' do have some value on the land both as soil particles and in a small amount of fertiliser value. However, despite the common sense of recycling these materials onto the field next door which would environmentally be much more sensible than any alternative, it is unlikely that the regulators will allow this in the near future. Sometimes the interpretation of 'the precautionary principle' becomes a stifling dead hand.

---

52. *Every Bottle of Prosecco May Erode 4.4 Kilograms of Italian Hillside*, <u>*New Scientist*</u>, January 21st 2019

This area of discussion will be very interesting to watch. It revolves around whether the arisings are seen as 'wastes' (a word with negative implications) or a 'resource' (a word which implies benefit and sustainability) i.e. not to be lightly lost or left un-used.

There is, in fact, an enormous amount of experience with deep drilling and putting the cuttings and spent fluids on a field close by – safely and under supervision by the UK Environment Agency. The agency people are well organised and always operate on the precautionary side. How might we add to that strength? Well, the most trusted, professional, soil scientists in the world are the members of the *British Society of Soil Science*. They could add credible technical knowledge and independent professional judgement to monitoring.

Would all of these things solve and eliminate all the risks? No, the world is not like that. However, we do have to make a choice. There is a real risk of a significant short-fall in power security and that might be very serious in consequences. It is a question of a balance of risks and sensible, common-sense, independent monitoring.

Traffic

There will certainly be heavy equipment brought in to develop a shale gas well. However, once set up, that equipment will depart, again on heavy trucks but when it is done, it is done. Where there is potentially a much bigger problem is the wastes generated from drilling the hole or well. Box one below gives a rather pessimistic view of how big the logistics situation could be if it is insisted that the waste arisings are trucked off site

---

Box 1

Background Logistics for UK development of Shale

Roughly speaking, around 1000 tonnes of cuttings and spent fluids come out of one drill hole. For a hole with multiple side laterals, the figure would be substantially more (but there would probably be fewer wells in a licenced area).

As an example, the Blackpool licensed area would need around 300 drillings to exploit it. That means *at least* 300,000 tonnes of cuttings and spent fluids.

For every 100 licenses issued, if of similar size, that would involve *at least* thirty million tonnes of output.

Truck movements: If that went into trucks for transport elsewhere, that would involve 8-wheelers of twenty tonnes nominal capacity each, i.e. 1.5 million truck movements into site and 1.5 million out, i.e. three million truck movements.

Trucking costs: If the typical trucking distance to dump were ten miles that involves thirty million truck miles. It is not just the financial cost of possibly around £100 million, it is the environmental and political cost of that movement which is likely to be critical.

Licenses: Nobody really knows yet how many licenses will be issued but it is quite possible that 200 or 300, possibly more, areas will be licensed. **It may end up with ten to thirty million truck movements, just to move the wastes. That is environmental and political dynamite**

However, there is a solution which has been tried and tested. A summary is shown in box two below. This was not actually fracking for shale gas, it was the construction of a gas pipeline from the mainland to the Isle of Wight. Fracking was not involved. However, the engineering of the drilling, the construction of the hole in the ground, and the fluids used in the drilling operation – all of these – could be used to sink the main well of a shale gas exploration and production well. So, yes, we do have the technology to at least do some of this very safely (more details on safety below).

Box 2

The Solution in a Case Study

IT HAS BEEN DONE with British technology!

A British designed and manufactured product has been successfully used to drill water wells and geothermal boreholes across the UK, and because of its environmental and performance credentials, can do the same for the shale gas industry.

This eco-friendly drilling product has been used very successfully on a major pipeline project, drilling deep under the Solent to put in two gas pipelines from the mainland to the Isle of Wight. This project was a world record in terms of distance and engineering challenge and avoided the requirement to trench across the Solent with the associated negative environmental impact.

4,500 tonnes of drilled cuttings and the spent eco-friendly drilling fluid from the drilling operation were spread to farmland at both ends, and the companies involved were able to prove agricultural benefit associated with the spreading of the spent fluid. This was "closed loop" recycling in its most efficient and environmentally beneficial form, carried out with permits from, and under the supervision of, the Environment Agency.

On the mainland, the cuttings were not loaded onto road-going trucks but, instead, loaded direct to large agricultural muck spreaders and spread direct onto the land of the Cadland Estate where the drilling took place. The estate farms manager, Richard Fellows, has now taken this sort of material and would "certainly be happy doing it again" provided the right products were used and the right permissions, advice and supervision were in place.

So, what's so friendly about this drilling fluid? Historically, nearly all drilling fluids were based on mineral oils. Those oils would be toxic if they leaked into aquifer and also if the cuttings from that hole were spread on land. So, they were dumped to landfill. The new eco-friendly "Pure-Bore®" from CSI is based on natural starch and is totally biodegradable and non-toxic, and provides an effective alternative to many of the toxic oil/synthetic based drilling fluids commonly supplied by the major oil-field service companies.

Environmental legislation certainly demands that risks of applying wastes to land are fully identified and properly managed. More than that, if 'wastes' are put to land, there must be an identifiable benefit.

In the case study above, these benefits were identified as in box three below and accepted by the Environment Agency.

Box 3

Why spread to Proximity land? Advantages forFarm or Forestry

From UK land, between 1.5 and 2.5 million tonnes of top soil are eroded by wind and rain into the sea every year. These cuttings will replace a bit of that and make the soil on the fields to which it is spread, a little deeper.

There is not usually a high level of plant nutrients in these cuttings but there is always some including phosphate (a major crop nutrient), Magnesium (important soil micro-organisms and to lactating mammals including humans) and mineral "trace elements". (Note that many of our soils are short of these minerals and that means the food we eat may be short; so these trace elements will help give us better food containing more trace elements so important to human health).

Reinstatement of Land

The well during construction is certainly active and very visible. Similarly, constructing pipelines below ground to take the gas away are a visible scar. However, reinstatement skills are very much alive with pipeline following and, provided there is professional skill employed and adequate monitoring, the land can be returned to original state fairly well with little cause for concern. Only a small and largely unobtrusive well head building will remain.

Production

Once a well has been drilled, commissioned and is producing, the output of gas per day will naturally fall. To keep it flowing, more high pressure fluids and probably sand will be pumped down the well to keep the cracks open and the gas flowing. The life of a well is difficult to predict at the start and highly variable. It might be none and it might be twenty or thirty years but not forever. This, then is a 'transition' fuel. It is a short life temporary expedient which has and will continue to change world politics and how the human race faces its future.

Conclusions

Can shale be done safely? Probably.

Can shale be done badly and dangerously? Of course.

The truth is that all technology is two edged. In the hands of good people with strong ethics and who do their research properly, any and all technology can be used for the benefit of mankind and the environment. Shale is no exception. Shale has been done badly and irresponsibly in some cases (certainly not all or anywhere near all) in the USA and, no doubt, elsewhere. It is up to governments and, most of all, the industry itself to deliver safe shale. I have long argued that, in the UK at least (and maybe globally) there is no organisation better placed than the BSSS, (British Society of Soil Science) to consider, supervise and advise on what goes down the hole and what happens to what comes out of the hole. Equally, I have no doubt that there are other organisations which could at least do that job adequately.

Done with the best available technology and ethical management, shale can be delivered in volume and with contained environmental cost. That environmental cost, we do know, can be very significantly lower than the burning of brown coal, black coal and oils simply because it is a clean burn. There is also the potential to use it in transport and significantly reduce urban pollution. It can and will continue to change the world economy and that can, if our politicians exercise their fair share of ethical management, provide for health care (preferably free at the point of need as in the UK National Health Service), care of and services for the old, water and food for those who need in in North Africa and Asia, and to make the human and wild life environment a better and more sustainable place to be.

If you have not seen and read the November 2015 issue of *National Geographic* magazine, maybe you should. Pages fourteen to fifteen argue convincingly that 2016 may be the tipping point in climate change.[53] At the time of writing, it is 2019 and, according to that article, we may already be in deep trouble. If you believe that if we can do something to protect ourselves and our children – we should. Then here is one solution.

Go for shale gas, right now, and spend the income on developing sustainable renewable technology and install it. Develop the gas production here in the UK and save imports. Spend the cash here to develop the technology and production right here and create the jobs right here. It is not just common sense, it is humanely and responsibly sensible.

We do have to make a choice. There is a potential risk of a significant short-fall in power security and that might be very serious in consequences; if and when, it is not just the lights that will go out. It is a question of a balance of risks and sensible, common-sense, independent monitoring.

---

53.   Kunzig, R: *This Year Could Be the Tipping Point, National Geographic Magazine*, Nov 2015, pp 14–15

Chapter 9: Sustainable Energy Production

Review

The problem with most forms of 'sustainable' energy sources is that they either need energy to be expended to extract or manufacture them in the first place, or need energy to be used to make a machine to use them. That original energy may well have come from burning fossilised fuels, so, until the system gets rolling enough to be able to use renewable fuels to provide that original energy, sustainable energy may be counter-productive. For example, the turbines to take energy from the wind, needs design, development, manufacture, transport, installation, commissioning, maintenance, decommissioning, all to be taken into account against the return. Fortunately, wind turbines do give a positive, full-life-cycle pay-back (see below) but some other alternative, 'green' sources of energy do not.

As already mentioned, as populations become wealthier, they almost always produce more waste. That is a two-edged sword; that uses resources including energy and (generally) hydrocarbon fuels, on one hand but on the other, 'wastes' can become resources. One of the new technologies is gasification of wastes to produce hydrocarbon fuels. For example, Tom Freyberg[54] reported in May 2015 Spanish firm Abengoa secured a $200 million contract from Fulcrum BioEnergy to build a bio refinery using gasification technology to convert 200,000 tons of municipal solid waste (MSW) into syncrude that will be upgraded into jet fuel

It is not just turning wastes into fuels that is happening, the discovery and production of shale gas and new oil finds will further alter the economics of energy and waste. It may be that waste to liquid fuel is uneconomic in itself, or will become that way, but is driven by political will and direction and/or subsidy.

It is the link between wastes and fuels, or better still, wastes and energy, which is a recurrent theme in this text and the survival of the human race. However, the clue to success lies in the production of carbon dioxide and the consequential locking up of oxygen. It is certainly true that carbon dioxide is a greenhouse gas related to global warming and we need to be concerned about that. However, our life depends on being able to breathe oxygen and locking large quantities up will also affect our lives. As this happens, we will gradually feel hotter, breathe faster, and get altitude sickness a bit lower down. And so will most of the life forms we know.

Reduction of Fossil Fuel Consumption

Ruminants are animals such as cows and antelopes, sheep and goats, buffalo and gnu, which have a rumen – a large stomach full of micro-organisms which allow these grazing animals to digest cellulose (grass and other green material). The process in the rumen is *anaerobic* which means that the micro-organisms do their work with little or no

---

54. Freyberg, T: *Gasification to Turn 200,000 tpa of Municipal Waste into Jet Fuel in Nevada*, <u>*Waste Management World*</u>, May 12th 2015

oxygen and, in the case of these animals, they produce methane, quite a bit of it, and belch it out into the atmosphere. Methane is a greenhouse gas with a far more potent effect than carbon dioxide. And there are billions of ruminants on the world's surface. There is some really exciting work being done by Australian agricultural scientists [55,56] to see if the micro-organisms in the stomachs of kangaroos (which eat and digest grass but do not produce methane) could be transferred in some way to ruminants. It has a good chance of success but, even if it does, it may take hundreds of years to change the world's population of ruminants.

There are many other contributors to climate change. It is not the purpose of this text to list and evaluate the relative contributions of each. What is already clear is that we have to tackle all of these as quickly as possible but it remains inescapable that burning fossilised fuels is a major contributor, most would argue, **the** major contributor to global warming.

The basic equations in managing real sustainability

Energy from the sun

Plants take CO_2 and water $\longrightarrow$ To make large Carbon molecules

$$CO_2 + H_2O \longrightarrow C_6H_{12}O_6$$

$\longleftarrow$ Animals and incineration push this the other way

The balanced chemical equation reads:

$$6\ CO_2 + 6\ H_2O \longrightarrow C_6H_{12}O_6 + 6O_2$$

Burning a small carbon molecule reverses this process:

$C_3H_8 + 5\ O_2\ 3CO_2 + H_2O$ - plus some energy as heat which we could use for making electricity

Propane   Oxygen

Burning a big Carbon molecule would read:

One molecule from petrol $2C_{36}H_{74} + 109\ O_2 \longrightarrow 72\ CO_2 + 74\ H_2O$

Rounded figures   1 tonne + 3.5 tonnes $\longrightarrow$ 3.2 tonnes + 1.3 tonnes

Never mind the Carbon dioxide, where is the Oxygen going?

The only reversal mechanism we have right now is the green leaf

Fig 9.1: The Basic Equations in Managaing Real Sustainability

55.   Chipperfield, M: *Kangaroo Bacteria Could Fight Climate Change, Telegraph*, December 6th 2007

56.   Klieve, A: *Investigation of the Microbial Metabolism of carbon Dioxide and Hydrogen in the Kangaroo Foregut* by Stable Isotope Probing, *ISME Journal* 8 (9) 1855-1865

Consider figure 9.1; plants take carbon dioxide out of the atmosphere and water through their roots and produce small sugar molecules. The most commonly quoted is the 6-Carbon sugar. When animals (including humans) use muscle energy, they push the equation the other way and we get the carbon dioxide and water back. Keep those two equations in balance and we have long term sustainability. The next line in the row of equations shows the burning of a small carbon molecule, propane, and how much of each element is used up.

The final equation is the burning of a large carbon chain molecule, actually found in petrol. This is what happens when a petrol-driven car is run.

Never mind the carbon dioxide, where is the oxygen going? The only reversal mechanism we have right now is the green leaf.

It is pretty clear that the burning of fossilised fuel reserves has to slow down and, preferably, actually stop as soon as possible, if not now. Tomorrow is too late. That, however, is not going to happen. Even if everyone agrees the principle (which they won't), they will then have to agree a timetable (which they won't), and then every nation will argue if this is a special case (which they will), and action will be too little too late (which it will be). Nevertheless, the fundamental need is to slow-down or stop burning fossilised fuel and this must happen and soon. We have to make a start. However, before turning to the look at developing Photosynthetic Carbon Capture and Storage (PCCS), it is important to put it in the context of a developing, wide range of alternatives, some of which are looked at below. By way of introduction, however, there is a problem.

The population of internal combustion engines. The starting point of all action is: we are where we are. It is sometimes said that there is a rat for every human on the face of the earth. Well, there is probably about one internal combustion engine for every human on the earth, too (more in developed countries). Whatever the relationship to global human population, global engine population is certainly a very large number. Whatever that number is, the industrial capacity available or conceivable must imply that it would take many years, several human generations, to build and commission a replacement of all those engines. Indeed, it is logical to assume that any solution might involve so much resource use so as to be counter-productive in, at least, the short term, i.e. push over the tipping point in climate change. The implication, logically, is that we have to use what we have got in the short run and that means use the existing engines. But logically, if we are to arrest global warming, we need to find a sustainable, liquid fuel to power them. That necessarily means biofuels. However, not all biofuels are sustainable. Looking first at some of the other alternatives, these, too, have difficulties.

Hydrogen

Doing some arithmetic, ref hydrogen versus petrol, produces results that might interest the reader, even if they have not yet bought a hydrogen-driven car. If we assume that the relative conversion efficiency of a petrol internal combustion engine versus a fuel cell and electric motor are similar, then the following figures may not be too misleading.

- The 'gasoline gallon equivalent' of hydrogen at one atmosphere is 357 cubic feet.
- That is, one US gallon of petrol will potentially release the same amount of energy as 357 cubic feet of hydrogen gas at one atmosphere pressure.

- There are approximately seven US gallons to the cubic foot.
- The density of petrol (gasoline) about 0.75
- The density of hydrogen at one atmosphere is 0.09 gm/litre.
- Hydrogen is frequently handled in bottles at 350 bar
- At 350 bar, that make 0.9 kilo per cubic foot

So 0.9 kilos of hydrogen can be stuffed into one cubic foot and have roughly the same energy content as one US gallon of petrol. In other words even at 350 bar the equivalent energy storage takes about eight times as much space as petrol.

Now add the heavy ironmongery (would aluminium bottles be strong enough in an accident?) needed to keep the gas contained. You can get to higher pressure but that means stronger bottles. Now add energy needed to get the gas into bottles at 350 bars. It won't be insignificant.

So even if you have a very lightweight chassis, the bottles capable of providing a range in the hundreds of miles are going to slow things up a bit. Maybe the answer is to generate the hydrogen very locally and store it in small bottles fitted with a very quick change device. So you go from one hydrogen point to the next. Frankly, it ain't going to work. QED

Clearly, some much better ideas than that are needed! One might be that methanol, ethanol, biodiesel and biogas are all relatively easily prepared locally from locally available plant material, are much easier to handle and can be used in certain types of fuel cell.

Hydro

Water Rams

If there is a good flow of water with useful volume, then the flow itself may be used to drive a pump, *kinetic* energy is the energy because of motion and mass. So water flow can be used to gradually push open a valve. Consider a mechanism allowing the flow of water and equipped with two valves. If water flow is used to push open the first valve and it gets to a designed-in limit and then snaps shut, there is quite likely to be a shock wave, or 'knocking' effect. The water flow may then start opening that first valve again, with the water flow going to waste. However, that shock wave from the opening and closing of the first valve can be used to open a second valve, pushing a little water through before it, in turn, snaps shut. In this way, the mechanism, known as a *water ram* can pump water up tens of meters and several kilometres. The mechanism is driven by water; there is no engine or electric motor, so it can be used in isolated places and is very low maintenance. Some may last over a hundred years. The mechanism needs a supply of water and good flow. Some streams can be relied on 365 days a year. If water supply is seasonal, a reservoir may be needed,

Water-Driven Turbines

There are two types of water-driven rotors.

Most countries in the world, throughout history, had water if there were wildlife or humans; water is one of the basic necessities of life. Most, but not all, those populations had flowing water. If there was flow, early on in human history, man harnessed that flowing water to produce mechanical power. Much of the early development of waterpower was rotary power, the flow of water over or past a water wheel or turbine, to pump water to irrigate crops, or to drive mills to grind grain to produce flour for food. Later some was used to drive industrial equipment. Later still, to drive electricity generators. Today, Canada, Brazil, Norway and Venezuela produce most of their internal production of electricity from hydro generation. They are able to do so because they have high mountains and rainfall (or snow) on to that high ground; water is up on high ground and can be used to drive a turbine on its way down and the seasonal and daily variation in flow of water, and hence power output, can in part be controlled by water storage behind dams. There is every reason why this can be done on a small scale, exploiting the same principles.

An over-driven wheel, with water going over the top of the wheel, usually a large one, uses only a little of the energy in the flow of water, using mainly the weight of water flowing into buckets down under gravity. The result is slow, high-torque rotation which suits stone-grinding of grain and driving mechanical machines.

The under-driven (or side-driven) turbine uses the kinetic energy of flowing water and, generally, has a high-speed, low torque output. This is especially useful for generating electricity.

The detail of building a hydro-electric generator is not within the purpose of this text. It is the purpose here to raise the possibility; if you have a source of water high up, it may be sustainable and 'free' power. It is one of the options and may be part of your 'energy mix'. There are many sources of information on this on the web at the touch of a button; one good one is *Homepower Magasine* which is a good source of up-to date, technical guidance and commercial.

Dams and reservoirs
Reservoirs often take much effort to make using a dam. It may be helpful to remember that dams can silt up. Therefore, it is a good idea to look at the sediment potential in the supply to the reservoir and, possibly, use a multi-stage set of dams, so that silt can be trapped and excavated by drag line from the first or early part of the system. Alternatively, a twin track sluice, or two parallel, first stage dams, will allow sediment trapping and relatively dry excavation of the one not in use for flow down to the main reservoir.

Research in the USA gives good, practical data on do-it-yourself (on the farm) reservoir building using old motor tyres as fill for dams and retaining banks. Stuart Hoenig in the USA has shown how car tyres can be used to make a dam or soil retaining wall.[57] Top soil is removed from the area to be built on, and a layer of tyres laid and pined there to form a base. That base can be built several meters high with each new layer of tyres roped with nylon or polypropylene to the tyre below. A fill of aggregate added to each layer will add the weight necessary to give the dam its weight. The tyres provide the cohesion.

---

57. Hoenig, SA: *The Use of Used Tyres in Water Systems*, USDA, ARS 173

Geothermal

The Earth is still very hot. The deeper, the hotter it gets. Drill down, pump water in and hotter water comes out. Drill deep enough and it is very hot water that comes out. For free! However, the cost of drill deep is significant. And by the way, deep geothermal, to work well, is always fracked (but nobody seems to remember that when talking about 'evils' of shale gas production).

An alternative is to lay a grid of pipes by mole plough, a bit like under-floor heating in reverse, and take low level heat from the ground. It takes a lot of ground and high temperatures are not reached but it works. It is possible to lay the pipe grid below cultivation depth and grow crops.

Biofuels from Crops Grown from Wastes

The ultimate eco-mimic of the carboniferous era!

Consider the following figures then follow the argument through. Take special note of the size of the carbon dioxide box, top centre, of each figure. Bear one other thing in mind: the individual figures are put here just as an example. It is quite possible to put different figures because of different crops, different yields, different views of how much carbon a particular material holds, and so on. What cannot be dismissed is the principle: this is a sustainable loop, it is the only one shown, to date, to deliver that sustainability which we seek. The evidence for that sweeping statement is simple and inescapable; it has been done before, 360 million years ago. The Carboniferous Era started that long ago and it took sixty million years to lay down our fossilised fuel reserves.

CROPS TO BIOFUELS - The Basic Route

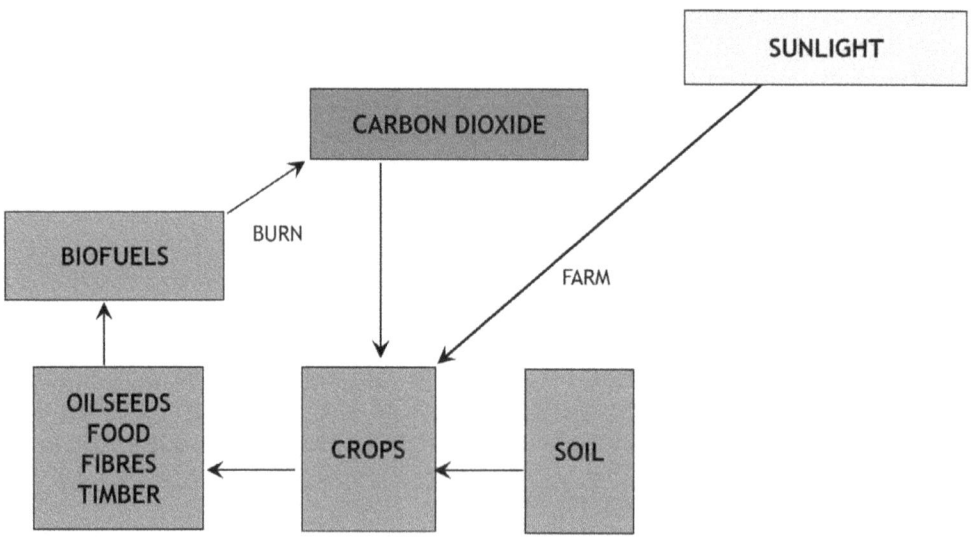

Fig 9.2: Crops to Biofuels – The Basic Route

Crops harvest sunlight by taking carbon dioxide out of the air and water via the roots and releasing oxygen. If the crop is burned in air, we use up the same amount of oxygen and release the same amount of carbon dioxide. On the face of it – a balanced and sustainable equation.

The last paragraph above stressed "on the face of it". The reality is that the situation in the figure above depends on where the crop inputs came from and, in particular, the fertiliser, and whether those inputs come from a sustainable route. Figure 9.3 below shows that route again but with the addition of mineral nitrogen fertiliser added to the diagram. Bear in mind that UN-sponsored research showed that one tonne of nitrogen nutrient produced in a typical modern USA factory takes 21,000 kWh of electrical power to manufacture and deliver. Note the difference in size of the carbon dioxide box at the top of each of the above figures. .Figure 9.4 shows a route which, in principle, is much more likely to be truly sustainable in that the plant nutrients come from waste and are already created. Whether the route is sustainable in practice depends on what resources are put into recycling. If a large amount of energy is put into processing such as shredding and transport distances are long, then that will reduce the chance of true sustainability. As ever, keep it simple and operate locally is likely to deliver energy efficiency. The carbon dioxide box at the top of the figure could be much smaller than in fig 9.3.

CROPS TO BIOFUELS - Mineral Fertiliser

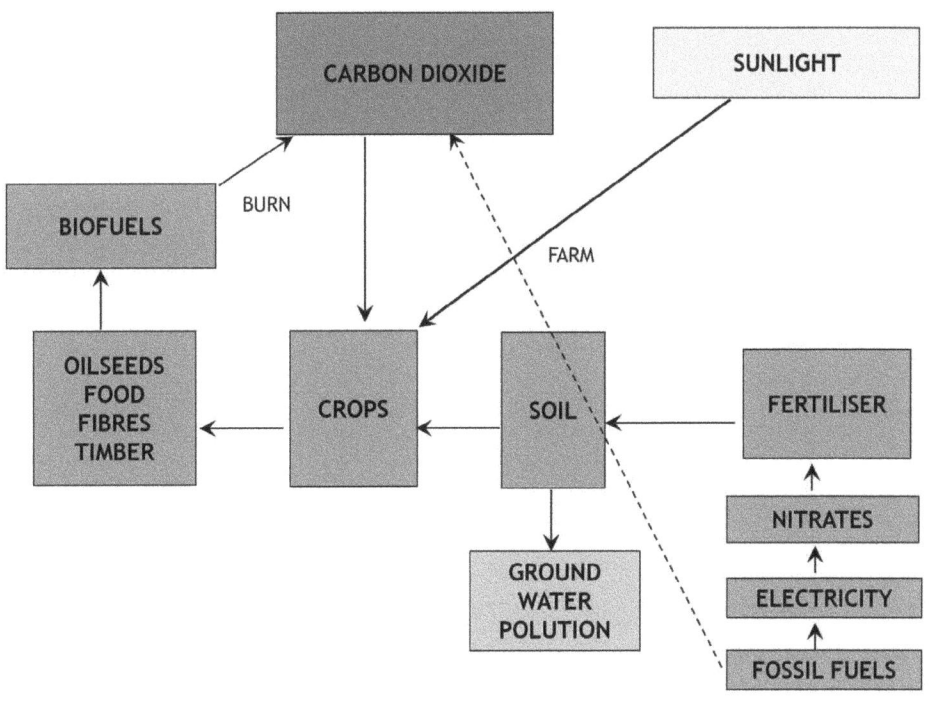

Fig 9.3: Crops to Biofuels – Mineral Fertiliser

One of the traps which biofuel production easily falls into is not to make sure that the feedstock it uses has an attractive carbon footprint. If the feedstock crop is grown with mineral fertiliser, it must be remembered that that fertiliser, where it is nitrogen fertiliser, will have been made by passing air through a large electric arc. The electricity is likely to have been made by burning fossilised fuels. Nevertheless to be fair, some manufacturers, have shown very significant improvements can be made in process efficiency and production of emissions. However, there is still a fundamental issue of inputs and sustainability.

Biofuels, from crops, from compost made from wastes is a different situation. It is not only carbon attractive; it is oxygen attractive, too. Also, when a crop produces oil seed which is used to produce the biofuels, the rest of the crop (the leaves, stems and roots) will help form a carbon sink with the carbon left in the compost.

CROPS TO BIOFUELS - From "Waste"

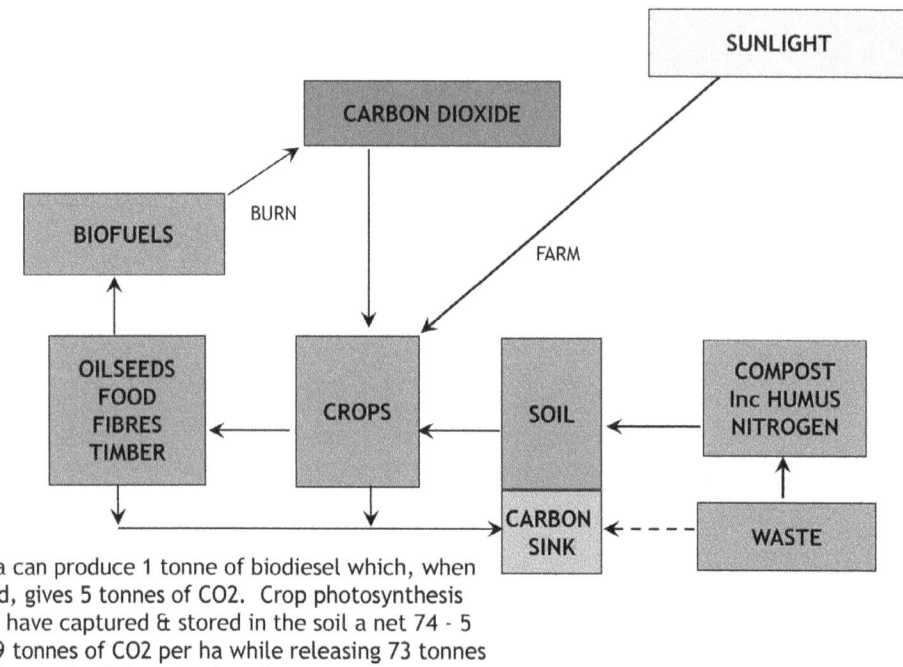

1 ha can produce 1 tonne of biodiesel which, when used, gives 5 tonnes of CO2. Crop photosynthesis will have captured & stored in the soil a net 74 - 5 = 69 tonnes of CO2 per ha while releasing 73 tonnes of Oxygen back for us to breath.

Fig 9.4: Crops to Biofuels – From "Waste"

Building the Carbon Sink

Much has already been made in this text of the opportunity to fix carbon and reverse global warming. It is also worth hammering on about biodiesel and PPO – Pure Plant Oil. Biofuel production from crops has a major global attraction *provided* it is done using wastes to fertilise the crops. It also, maybe just as important, can be done on a large or small scale and done locally. It is possible to limit trucking to collect wastes on a proximity basis, grow and harvest oil-bearing crops in that locality, convert to biodiesel on the same local operation and only export the surplus diesel production. If the whole operation is carried out on a community basis, the energy equations do stack up and there is a real gain in carbon capture. It is important to observe, however, that the key to success is not CCS (Carbon Capture and Storage – which is being talked about with respect to limiting the damage caused by burning oil and coal) but in PCCSS (Photosynthetic Carbon Capture and Storage in Soils) using photosynthesis and soil. If PCCSS can be used while significantly cutting the burning of fossilised fuels, then there is a very distinct possibility of reducing and reversing global warming.[58]

Plants harvest sunlight. A plant uses the energy from the sun, with the help of the chlorophyll in the green leaf, to take water (hydrogen and oxygen) via its roots and carbon dioxide (carbon and oxygen) via its leaves from the air to make sugars and then oils. There is a bonus because there is some oxygen left over and that goes back into the atmosphere. In a balanced ecological system, plants and animals circulate oxygen and carbon dioxide in this way. If we chose oil-producing crops, then we can use that oil to produce biodiesel. If the process can be done locally, then the logistics and energy consumption involved in trucking to centralised processing and distribution can be dramatically cut. Figures from decentralised operations in the UK[59] indicate that this reduction in tonne-truck-kilometres (used as an indicator of the energy cost of taking one tonne on a truck for a kilometre (if it is preferred to be imperial, tonne truck miles can be used) can be in the region of 65 to 85%. If the crops can be grown with fertilisers made from wastes (usually by composting and/or mulching), then it is possible to eliminate the energy involved in industrial manufacture of mineral nitrogen fertilisers which use significant amounts of electrical power (usually generated by burning fossilised fuels). There are many local wastes which can be used to produce a choice from over a hundred oil-producing crops.

The Carbon Dioxide Figures

The figures given here in fig 9.5 below and in this text are from a UK research and development project[60] but similar figures will apply elsewhere and with other crops. These figures are related to oil seed rape and the wastes are separated municipal biodegradable wastes, MDF from furniture production/recycling and some industrial wastes. See figure 9.5 with the explanation which follows. The figures used are an actual field example which might be duplicable elsewhere and are a reasonable guide. It is quite possible to argue with the detail of the figures but not the principle.

---

58. Butterworth, B: *Reversing Global Warming for Profit*, *MX Publishing*, London, 2009

59. Empirical experience in the *Land Network* consortium of farmers who recycle wastes to land. Some of this experience is published, much is held by Land Research Ltd.

60. Butterworth, B: *How to Make On-Farm Composting Work*, *MX Publishing*, London, 2009

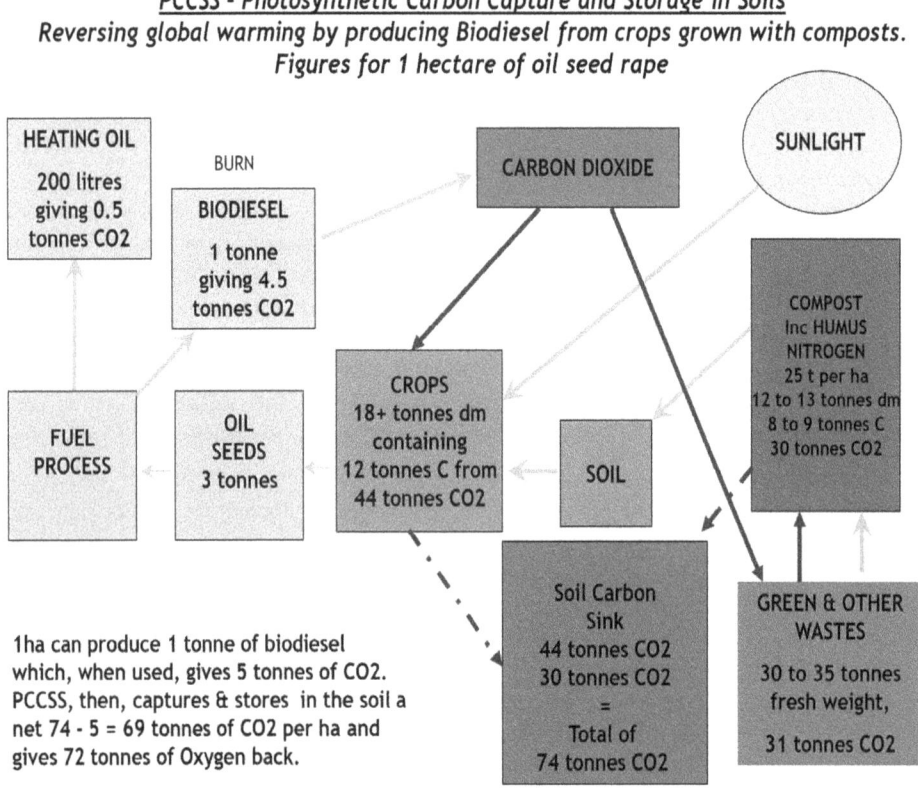

Fig 9.5: Photosynthetic Carbon Capture and Storage in Soils

The ultimate eco-mimic of the Carboniferous Era

One hectare of oil seed rape crop will produce around three tonnes of oil seed rape and that will yield about one tonne of biodiesel.[61] In producing that three tonnes of seed, the crop will also have produced probably more than five to ten tonnes (say typically 7.5 as an example) of leaf and stem. Generally, most crops will produce as much dry matter below the ground as they do above the ground. Therefore, the total dry matter production might be in the region of eighteen tonnes per hectare. Now, that eighteen tonnes of dry matter is mainly carbon and will contain about twelve tonnes of carbon. That carbon will have been taken out of the atmosphere by the green leaf chlorophyll process. As the atomic weight of carbon is twelve and that of oxygen is sixteen, and there are two oxygen atoms to each carbon in carbon dioxide, the original twelve tonnes of carbon in the dry matter will have removed forty-four (12+16+16) tonnes of carbon dioxide out of the atmosphere. However, that is not the end of the capture (remember that forty-four and carry it forward).

---

61.   Empirical experience in the *Land Network* consortium of farmers who recycle wastes to land. Some of this experience is published, much is held by Land Research Ltd.

If the fertilisers are made from biodegradable wastes, then those wastes also contain carbon which will have come from the same green leaf process. Compost made from green wastes would, in the (UK) Land Research Ltd R&D programme, be likely to lose around a quarter of its weight (almost all water) during the process. Regulation in the UK will allow compost to be applied to the crop at around twenty-five fresh weight tonnes per ha. To produce that twenty-five tonnes would take maybe thirty to thirty-five tonnes of green waste which would have taken around thirty-one tonnes of carbon dioxide from the atmosphere.

The twenty-five tonnes of compost used on each hectare would contain probably twelve tonnes of dry matter and, therefore, around eight tonnes of carbon, which would have been fixed by removing a bit more than thirty tonnes of carbon dioxide out of the atmosphere. In fact, the original green waste would have had a little more carbon in it, say thirty-one tonnes, but a little is lost back to carbon dioxide released to the atmosphere by the composting process. So, the total removal of carbon dioxide from the atmosphere is the thirty tonnes in the compost which is locked up into the soil *carbon sink*, plus the forty-four tonnes in the crop. That makes this programme remove a net seventy-four tonnes of carbon dioxide from the atmosphere to produce one tonne of biodiesel. When the biodiesel is burned, and its co-product bioglycerol which will also be produced by the biodiesel production process, there will be five tonnes of carbon dioxide released back to the atmosphere. The net, then, is seventy-four minus the five equals sixty-nine.[62]

Global Warming Reversal

The conclusion, taking it all into account, is that this process, on these figures, will remove a net of sixty-nine tonnes of carbon dioxide from the atmosphere and release around seventy-three tonnes of oxygen back for us to breathe. The reverse would have happened if this material had been burned in an EfW (Energy from Waste) plant, producing carbon dioxide and removing oxygen from the atmosphere. EfW processes incinerates waste to produce heat, which can be used to produce steam to drive a generator, to produce electricity. However, if the figure 9.5 process were used with the same waste, through the green leaf, burning the oilseed rape oil in a diesel engine, direct coupled to a generator would produce more electricity *plus* the sixty-nine tonnes of carbon dioxide removed from the atmosphere and seventy-three tonnes of oxygen put back in. There is no way EfW can get anywhere near these figures. What EfW would have done would be to turn the carbon in the wastes back to carbon dioxide – back to the 31 tonnes right back at the beginning of this discussion. Comparing these as alternatives, the compost route is 69+31=100 tonnes better at removing carbon dioxide, gives us back the oxygen and delivers more energy. This route really is sustainable. The detail of the figures can be argued about but they are not fundamentally misleading. The principle is sound.

Clever? Not really. It is just a question of mimicking what we observe in nature and how it managed to create ecological balance in carbon management. In the UK, there has, at peak, been around 450,000 hectares of *set-aside* land under EU rules on limiting arable production. When that was set up, nothing could be grown on that land. After some years, the EU, in its great wisdom, allowed to grow energy crops on that land.

62.   Butterworth, B: *Reversing Global Warming for Profit*, MX Publishing, London, 2009

Later still, set-aside was dropped. At 44 tonnes of PCCSS (*Photosynthetic Carbon Capture and Storage in Soils*) that land, could have locked up, every year, over twenty million tonnes of carbon dioxide (24% of the *Kyoto Protocol* estimate of total UK emissions) and produced 450 million litres of biodiesel with less than one million tonnes of carbon dioxide produced when actually used.[63]

The figures above show how effective this can be and suggest we could reverse global warming.

Well, it is a bit more complicated than these figures suggest, of course. Firstly, there is energy used up in process and logistics. This energy use can be reduced if the process is based on community operation. Secondly, there is leakage of carbon dioxide from PCCSS systems at maybe 20% in conventional cultivation systems and maybe as little as 10% in reduced cultivation or 'zero tillage' (called 'direct drilling' in the UK) systems. Reduced cultivation marginally increases the emission of another greenhouse gas, nitrous oxide. However, there is a balancing factor which is that the method described here saves very large amounts of energy which is used and transporting mineral fertilisers which consumes very large amounts of electricity (remember the figure of 21,000 kWh per tonne of mineral nitrogen fertiliser – and that is in a modern factory). So, these figures are a simplification for the sake of presentation here. However, they are not misleading. This could be the delivery mechanism, maybe it is the *only* delivery mechanism, which will work. This is a global scale system which already exists.

Technology can deliver management solutions but only if there is a political will to provide the wider framework to accommodate a raft of measures. Firstly we have to dramatically cut fossilised fuel burning and quickly. We need to nurture those reserves. One idea here is to fund the United Nations by using it to administer a global aviation fuel tax and we need the USA and China to seriously sign up to Kyoto principles. Fortunately, under President Obama, the USA tide did turn but, unfortunately, the worst possible thing happened when President Trump put the brakes on. Secondly, we have to organise global waste management to re-direct it into safe land management for energy. Thirdly, we have to study and manage the effects of the potential loss of global dimming. Lastly, and by no means least, we have to consider that much of the cash which could drive the political framework to make these things come right can only come from getting the oil companies to buy into the framework. They could start by thinking about looking at the centralised production of bio-ethanol and about distribution of that bioethanol and the surplus of biodiesel from community production from wastes. Delivery is possible but the clock is ticking faster than ever. Remember that the argument here depends on producing those biofuels from crops that are fertilised not from mineral fertilisers manufactured specifically for the purpose, but from wastes.

Wind

Put the negatives first. At a national level in terms of a state strategy, it is as well to remember that there is a strategic issue in what happens when the wind stops blowing. It is clear that the alternative sources (power stations fired with other fuels) cannot be shut down; so there cannot be a capital saving and the wind power installation, therefore, has to be charged with the capital cost of the power stations it replaces when the wind blows (but which cannot be shut down when it does not). The truth is that wind turbines

63.   Butterworth, B: *Reversing Global Warming*, *ReFocus* September/October 2006

are unlikely to be economically justified and sometimes not environmentally justified, unless really well designed, reliable equipment is installed in a consistently windy place. Having said that, in the right circumstances, it can be both economically and financially sustainable.

In an analysis by staff at Stamford University, USA, wind turbines gave clearly the lowest life cycle energy cost per kWh of electricity output compared with photovoltaic and solar-thermal.[64] Looking at energy return, or *energy payback*, one study showed[65] a payback over the expected life of a particular wind farm at twenty-one, i.e. lifetime production of power was twenty-one times what was put in to make it work over its lifetime. One of the more generally valuable studies was a meta-analysis by Ida Kubiszewski *et al* involving 119 turbines from fifty different studies which showed an EROI of around twenty, i.e. the *Energy Return on Investment* was that the energy output over the life of the turbines was twenty times the total, life-time energy cost including decommissioning.[66] So, generally speaking, wind turbines are the best we have yet in wide commercial use for giving back more than we put in.

There is one twist in the economics which may underline the real truth; it may be that an installation is financially justified by paying for it to be installed at today's values and justifying it at future energy prices. It may also be a way of strategically securing energy supplies or, at least, making a nation less insecure in the face of international energy power politics.

The rule, as ever, is to do the homework and tell the truth at all stages. Again, for the equipment, there are plenty of commercial sources to be found on the web.

Solar

There are two sorts of solar energy collectors. The first uses the sun to heat water in a thin jacket, i.e. 'solar thermal'. In a hot, sunshine climate, because of few moving parts, this system may well be economically and environmentally sustainable. In less sunny climates, this may not be the case and the arguments applied to wind power are then part of the decision making process.

The second mechanism is to use a photo-sensitive device to convert sunlight to electrical power directly, i.e. PV or photovoltaic. Putting the negatives first, it is probably true that if all the PV panels so far installed were added together to give the total life-time energy put in, then it might not be replaced by what comes out over a twenty-five year life. Much of the equipment, historically at least, costs more in energy to make it than will ever be produced out of it. However, the latest technology, with well researched sites, properly maintained can be efficient and productive.

64.   Dale, M: *A Comparative Analysis of Energy Costs of Photovoltaic, Solar Thermal, and Wind Electricity Generation Technologies*, Stanford University

65.   Rajuel, M & Tinjum, JM: *Life Cycle Assessment of Energy Balance of a Wind Energy Plant*, Geotechnical and Geological Engineering, March 20th 2013

66.   Kubiszewski, I, Cleveland, CJ & Endres, PK: *Meta-Analysis of Net Energy Return for Wind Power Systems*, Renewable Energy Volume 35 Issue 1, January 2010

While, in the short run, there certainly have been many devices which were very inefficient and not environmentally attractive, PV is likely to be really attractive in the long run especially if low cost batteries become available for local storage to give 24/7 power supply. We will get, increasingly as time goes on, more efficient devices which are cheaper and more efficient. The rules remain the same; homework and look for independent and credible research to justify the type and brand.

There is at least some research which gives a useful guide. Much of that research suggests an energy payback period, *on the panels themselves*, of two to ten years. This research did not include the costs of logistics (storage and transport), erection and supporting framework cost, commissioning, maintenance and decommissioning.) The US Department of Energy quotes in one of its published documents several pieces of research from all over the world where researchers of some academic standing looked at the question and showed a wide range of results but they were all presented by the department as positive. The summary of research indicated that crystalline modules were significantly better than a few years ago and their efficiency would improve still more. The conclusions were also that thin-film technology was currently more efficient than crystalline and would continue to keep its lead over crystallines. Further, as new technologies develop, this is very likely to increase efficiencies again. For example, the sun emits a much wider range of energy wavelengths than just the visible spectrum. New panel technologies will collect infrared and ultra violet and maybe wider.

Nevertheless, reading between the lines, most of the quoted research appeared to be limited to the energy cost of manufacture of the panels themselves. There was limited or no indication that the total energy cost of a working installation had been taken into account. One of the bits of research even admitted that they had not even included the thin frame that surrounds the panel before putting into its packaging. So, there is some doubt about whether the research figures cover all the energy costs, not just of manufacture, but also of packaging, shipping, land transport, installation, site infrastructure, site works, supporting frame construction, commissioning, maintenance, cleaning and failures, and decommissioning.

A Note about Pay-Back
Does energy pay-back matter? Maybe not. It depends on where you sit. If it makes money for everyone involved; maybe that is good enough? If it takes government subsidy to make it economic, only the taxpayer might not agree. One thing 'renewable' (but not necessarily sustainable) might be seen to deliver is less reliance in future on imported oil, gas and electricity (Yes, the UK imports significant amounts of electricity from France). There is another plus; whatever the energy cost, it is paid for at today's cost. The energy produced over the life of the installation pays back at tomorrow's energy values; it is an investment in tomorrow.

There is a negative here in that the researchers generally were interested in energy payback and money. What we need with all these forms of renewable energy sources is life-time carbon dioxide saving. These figures are more difficult to get.

Leaving solar on a positive, during late 2018 and early 2019, the idea of creating the roof on a building from solar panels, without building the roof first, began to be a commercial reality (see www.landreseaarchonline.com and scroll down to January 18th 2019). At least, someone has finally grasped that making a conventional tiled roof, and then retro-fitting solar panels is little short of insane suicide. We have the technology to make a roof out of solar panels and we should never, from this day on, ever make another conventional roof.

EfW – Energy from Waste

Pyrolysis
Basically, these processes use heat to raise the temperature of feedstocks so far as to turn them, or part of them, into gas. Many wastes contain large organic carbon molecules and heating them in the absence of air produces methane which can be used to drive a generator and produce electricity. Waste heat from the engine can be used, at least in part, to raise the temperature of the next lot of feedstock. Again, methane is a clean burn producing carbon dioxide and very little else. However, methane may not be the only gas. Usually, if not always depending on the feedstock, there will be a residual ash which may have a significant amount of heavy metals.

Incineration
Burning waste in air, incineration, is common and has a number of issues. Firstly, comparatively little heat (net energy) comes out of the process. Secondly, burning in air produces carbon dioxide which, in terms of what we have to face in terms of global warming is, to put it politely, counter-productive. Thirdly, the flue gases may be very dangerous in practice. Fourthly, there is a significant ash which is likely to be a problem to get rid of because of not only heavy metals but other carbon-based residues, some of which may be quite dangerous environmentally, (thus necessitating specialist landfill disposal.)

Both of these routes turn organic carbon into carbon dioxide. There is another route and that is to find a way to separate the organic carbon and compost it to produce soil humus, ie, a bio-active carbon contribution to the global sink (see below).

Wave Power
Wave power is, like wind, for free and, like wind power, is OK when the waves are there and not OK when they are calm. Nevertheless, here is a potentially sustainable source of non-polluting power.

Tidal
The fact that the moon's orbit of the Earth twice a day moves trillions of tonnes of seawater back and forth every day is pretty exciting and it is for free. It is true that people living near the coast or estuaries have harnessed the tide for centuries – but not very much on a big scale. Why not? Well, the financial world has had this long love affair with oil and the motor car. Even now, in the face of a crisis, there is a blind pursuit of wealth from oil to the active suppression of other, more sustainable, energy sources, including power from the sea. The sea is big and can be bad-tempered. Nevertheless, there is enormous un-tapped potential here and that applies to both tide and currents in the sea (also see www.landresearchonline.com and scroll down to January 22nd 2019).

Microbial

Now, this is potentially an interesting possibility. It has been known for some years that micro-organisms can digest organic carbon and produce electricity. There has been some progress recently on processing wastes to produce environmentally safe effluents and generating a surplus of electricity at the same time.[67] The big plus here is the possibility of treating waste, including domestic sewage, at the point of production, so as to avoid transporting somewhere else, and delivering electricity.

---

67. Adee, S: *Bacteria Made to Turn Sewage into Clean Water – and Electricity*, *New Scientist*, July 27th 2017

Chapter 10: Sustainable Oceans

Probably the nearest we currently get to claimed sustainable oceans is fish farming. The problem with most such farms is that broadcast feeding develops significant pollution underneath the cages.

In principle, we ought to be able to use the sea to produce algae and, from them, to produce renewable liquid biofuels. The problem is tide control and extremes of weather. Various estimates have been made of how much land would be necessary to produce enough biofuels for the world's cars, trucks or aeroplanes. Two thirds of the globe is covered by sea and we are a long way off managing that with green algae production and the potential management risks of using this route are very significant in terms of pollution risk simply because of tides, winds, currents and water turbulence.

Developing using the oceans in the time we have to solve the triad problem is not likely – it is not long enough. However, this does not mean that we should not stop polluting the oceans and try to clean them up. The oceans are still part of the carbon dioxide in the atmosphere picture. If we kill off the oceans, we will kill everything else off, too.

Section 4

Building BACS – The Global Bio-Active Carbon Sink

Chapter 11: Soils and Greenhouse Gases

Photosynthesis

As discussed in the review at the start of *Chapter 9*, plants with green leaves can undo the problem created by burning fossilised fuels.

Burning one of the molecules in petrol reads:

$$2C_{36}H_{74} + 109O_2 \rightarrow 72CO_2 + 74H_2O$$

Note the loss of oxygen and then read the next equation which shows one of the steps in what photosynthesis does inside the green leaf, using sunlight as the energy source to drive the equation:

$$6CO_2 + 6H_2O \rightarrow C_6H_{12}O_6 + 6O_2$$

...or, if we relate this second equation's amount of carbon dioxide to that in the first equation:

$$72CO_2 + 74H_2O \rightarrow 2C_{36}H_{74} + 109O_2$$

Note that there are two products; a 6-Carbon sugar and we got the oxygen back. The sugar is most likely to be built into cellulose and a range of other valuable components including starch, dietary fibre and plant protein.

Despite this rosy picture, plant breeders have been trying, ever since man began to cultivate crops, to get higher yields. Part of that striving has been cultivations to control weeds, spay chemicals to control all pests, fertilisers and plant breeding. Part of the plant breeding has been to change the way the plant structures itself to catch more sunlight. For example, breeders of maize have managed to structure the upper leaves to grow to such an angle that more light goes through to the lower leaves. There has, however, been an assumption that we are stuck with photosynthesis and its efficiency as originally evolved, as the process which, despite significant effort from very good scientists has not, until recently, been changed. There is some breakthrough research that shows that the current photosynthesis process was evolved when the world was different.

When green plants evolved, billions of years ago, the atmosphere had much less oxygen in it. The photosynthesis process uses an enzyme which hooked onto the carbon dioxide without risk of hooking up with oxygen by mistake. As the atmosphere changed and the content of oxygen changed, the mechanism was confused and increasingly took up oxygen instead of carbon dioxide.

Amanda Cavanagh, University of Illinois, is working on this. She has worked on genetic re-engineering the photosynthesis pathway and succeeded in a 40% increase in yield in tobacco. Her team is now working on crops such as cereals and soya. The same report in *New Scientist*[68] says that other teams are working on improving the efficiency of photosynthesis by other routes, with some success. The USA government alone is spending hundreds of millions of US Dollars in supporting the development of 'synthetic'

---

68. Cavanagh, A: *Evolution's Biggest Mistake*, *New Scientist*, January 12th 2019, p16

photosynthesis, i.e. outside the green leaf, in the laboratory and in factories. Yet again, this is very hopeful but, yet again, by the time it could have a significant effect on global warming and food production, it will be too late. This does not mean that we should not be doing that research and pressing on with developing its results, more that we should be re-doubling our efforts to limit the triad of problems here and now.

Nat Lewis from the California Institute of Technology, Pasadena is quoted by the *New Scientist* as saying "Artificial photosynthesising systems must be three things: efficient, cheap and robust. Cracking all three at once is the problem". There is reason to believe that the efficiency of photosynthesis by the green leaf can be significantly improved and this will be a major contributor to feeding the global population. In that development, the process will remove more carbon dioxide from the atmosphere and pump oxygen back in – so making a contribution to stabilising the atmosphere and reversing the global warming process.

Organic Carbon in Soils

In farming, there are two key issues related to global warming. Firstly, the green plant can take carbon dioxide out of the atmosphere and pump oxygen back in. Secondly, the carbon-based organic matter in soils does oxidise (at varying rates) and 'leak' the carbon back as carbon dioxide or, worse still, methane, and the nitrogen back as nitrous oxide (another, very potent greenhouse gas with over 300 times to warming effect as carbon dioxide). However, these leakages back into the system are relatively very small and can be controlled to some extent. As Sara Wright's researches showed,[69] and others before and since the rate of leakage is dictated by a number of factors including soil temperatures but maybe the major one is under the farmer's control and that is the violence of the cultivation. High power input generally increases the rate of oxidation of organic matter, producing both carbon dioxide and nitrous oxide. The greatest effect results from the use of PTO-driven implements to 'force' a tilth ('PTO' means power-take-off and applies to implements which use the power of the tractor engine in a direct drive to rotate or oscillate implements to force a tilth). Research results show a wide range of emissions, from as much as 1.5% to less than 0.5%, of the total nitrogen back to atmosphere as nitrous oxide. Conventional cultivation systems, involving implements such as the mouldboard plough, tined cultivators and power harrows, in several passes, will cause relatively rapid oxidation of organic matter and maybe 35% per annum of the carbon in the organic matter will be lost on a declining basis,[70] i.e. 35% of what's left will be lost next year, and so on. However, with direct drilling, or zero till, that figure may drop to only a 10% loss[71,72] sometimes less.

69. Sara F Wright & Kristine A Nichols are with the SDAARS. This research is part of *Soil Resource Management* and *ARS National Program* (#202)

70. Empirical experience in the *Land Network* consortium of farmers who recycle wastes to land. Some of this experience is published, much is held by Land Research Ltd.

71. DTI unpublished reports by Land Network International Ltd under the Enterprise Initiative Programme

72. ICI Plant Protection, as was, pursued their market for Gramoxone with both commercial (including Land Network International Ltd) and academic enthusiasm in the 1980s with many publications and support trials, including by universities

Greenhouse Gases

Nitrous Oxide

One of the complicating factors in all of this is that carbon dioxide is not the only greenhouse gas produced in natural process including in the soil and in composting processes. Nitrous oxide is one of these and it is many times (300 times) more potent in terms of global warming than carbon dioxide.

The EPA of the USA says[73] that nitrous oxide is emitted when people add nitrogen to the soil through the use of synthetic fertilizers. Agricultural soil management is the largest source of N2O emissions in the United States, accounting for about 74% of total US N2O emissions in 2013. Here, again, it may be possible to argue about the exact figures in any particular situation, but not about the basic principle. Manufactured, 'artificial' or mineral nitrogen fertilisers do release nitrous oxide and, from a greenhouse gas point of view, that is not desirable. The report went on to say that nitrous oxide is also emitted during the breakdown of nitrogen in livestock manure and urine, which contributed to 5% of N2O emissions in 2013. There is a partial answer to this leakage and that is to compost the manures with other, high-carbon materials such as straw or municipal green waste.

It is important to note that nitrous oxide will also be emitted from organic soil systems and from the composting process. However, emissions of this gas are generally very significantly less from organic-based systems than from operations based on mineral fertiliser and other highly soluble fertiliser systems such as the use of liquor from AD (anaerobic digestion) processes.

Methane

Undisturbed soils will naturally emit small amounts of carbon dioxide and some nitrous oxide. Soils will also release some methane particularly if fertilised with mineral nitrogen, These amounts increase with cultivation.

---

73.   *Overview of Greenhouse Gases – Nitrous Oxide, US EPA*

Chapter 12: How *The Closed Loop* Really Works
...and Why Natural Ecosystems Don't 'Leak' Nutrients or Self-Pollute

The processing capability of the soil dwarfs human industry. An American scientist once calculated that the micro-organisms in an acre of arable soil would weigh as much as a fully grown cow. These are microscopic, tiny organisms of nano-scale (*nano* means ten to the minus nine, e.g. one nanometre is one 100,000,000[th] of a metre). Not just millions, not billions, but trillions of micro-organisms[74] in just a handful of soil, form a dynamic universe with enormous processing ability. Multiplication rates and biodiversity are enormous by human standards, and so is the range of their appetites. It is generally true that, given sufficient time, nature will deal with any material which has been spread out thinly enough, and bring the system back to a dynamic, balanced 'normality' (where 'normal' means sustainable).

Much of the material in this section on nitrate pollution was first published in papers published in *Resource*, journal of the American Society of Agricultural and Biological Engineers[75] and in *Landwards*, the journal of the British Institution of Agricultural Engineers in the early summer issue of 2002.[76]

Understanding the mechanisms in what is commonly called *the closed loop* and managing those mechanisms makes recycling to land dramatically safer in environmental terms. The figures below show the principles of the closed loop. Organic materials, and inorganic ones which have food value for the soil micro-organisms, do *NOT* break down directly to form humus. Such materials added as 'waste' are consumed by micro-organisms and turned into their own bodies. It is the breakdown of these bodies which form the complex, relatively stable black tarry material which gives soils their dark colour, generally termed *humus*.[77] So, knowing how to feed these organisms is the first step in the management of composting wastes and of the soil. It is also important to see that the compost heap and the soil are not separate operations. Mostly, everything that goes on in a compost heap would also happen in the soil, even pathogen destruction. The big advantages to farming of composting before spreading to land is to use the temperature to kill weed seed and to fit the logistics of the farm cropping year. The micro-organisms feed, multiply and die, then break down into humus. Humus is an extremely complex mixture of heavy molecules of hydrocarbons (the same process which makes crude oil), carbohydrates and proteins (which lock up the nitrogen). These molecules are large

---

74. Kinsey, N: *Hands on Agronomy*, <u>Acres USA</u>, 1999

75. Butterworth, B: *Clamping Down on Compost, Resource, <u>American Society of Agricultural and Biological Engineers</u>*, April 2006

76. Butterworth, B: *Nitrate Nonsense, Landwards, <u>Institution of Agricultural Engineers</u>,* Summer 2002

77. Jeffries, P: *The Contribution of Mycorrhizal Fungi in Sustainable Maintenance of Plant Health and Fertility. <u>Biology and Fertility of Soils</u>*, 2003 Vol 37 pp1–16

and insoluble and there is no limit to the quantities that can be put onto the soil safely. The evidence for this can be found in any natural ecosystem such as the Fens in the eastern part of England. When the Dutch engineer Vermuyden drained them nearly three hundred years ago, some were ten to fifteen metres deep. Up until now, farmers could grow crops there every year (for three hundred years!) exporting the harvested products with the nutrients they contained, including the nitrogen, and *never* need to add any fertiliser. Clearly, there had been an enormous reservoir of crop nutrients but the Norfolk Broads are not polluted with green slime and dead fish. Not all nitrogen is the same! NVZ's are 'nitrate vulnerable zones' and are the UK government's attempt to create regulations which will prevent nitrate pollution of groundwater. Incidentally, the evidence is that these zones have, at best, limited value, maybe none. If the mineral nitrogen fertiliser were added to a compost heap and processed into organic humus there before spreading, significant losses to groundwater would be eliminated.

Those large molecules of organic matter as humus will remain for centuries until long strands of soil fungi, called mycorrhiza, linked at one end to plants requiring food, start consuming them. These mycorrhiza either go up to and envelop the plant root hair, rather like the placenta in a baby mammal in the womb, or actually cross the root hair wall into the plant.

This is a closed conduit and why natural ecosystems do not lose their nutrients (as do in systems based on soluble fertilisers) and do not leak enough nutrients to pollute themselves or groundwater.

As common sense might indicate, the green-leaved plants and the soil fungi evolved together over millions of years and they operate at the same soil temperatures, so the system is demand-led. This system is how all natural ecosystems not only eliminate nitrate pollution, they eliminate all such out of balance pollution including phosphates, potash etc, etc. There is a further advantage, as the system locks up carbon in the soil. The 100 million tonnes of 'waste' produced in the UK per annum and which could be recycled to land would, if incinerated, produce around 75 million tonnes of carbon dioxide per annum which is 10% of the *Kyoto Protocol* estimate of total UK emissions. Composting to land can lock that up. Nevertheless, regulation in the EU, and UK in particular, is progressively limiting recycling to land and increasingly incinerating wastes under the guise of *Energy from Waste* – EfW. As discussed above (see *Section 2*) EfW is really little more than incineration, producing GHGs.

The figures on the following pages[78,79,80] show how natural ecosystems manage to 'leak' enough, and only enough, to keep the system working without pollution or starvation, i.e. in balance or 'sustainably'. Figure 12.1 shows a conventional view of how the system works.

The conventional view of how plants feed with the assumption that nutrients get to the plant via solution in the groundwater. With mineral fertilisers, this is probably either partly or completely true

78. McKenzie, D: *Facing the Resistance, New Scientist*, January 19th 2019, pp20–21

79. Rose, SC et al: *The Design of a Pesticide and Washdown Facility, British Crop Protection Council Symposium*, November 2001

80. Butterworth, B: *Survival – Sustainable Energy, Wastes, Shale Gas, and The Land, Amazon*, 2016

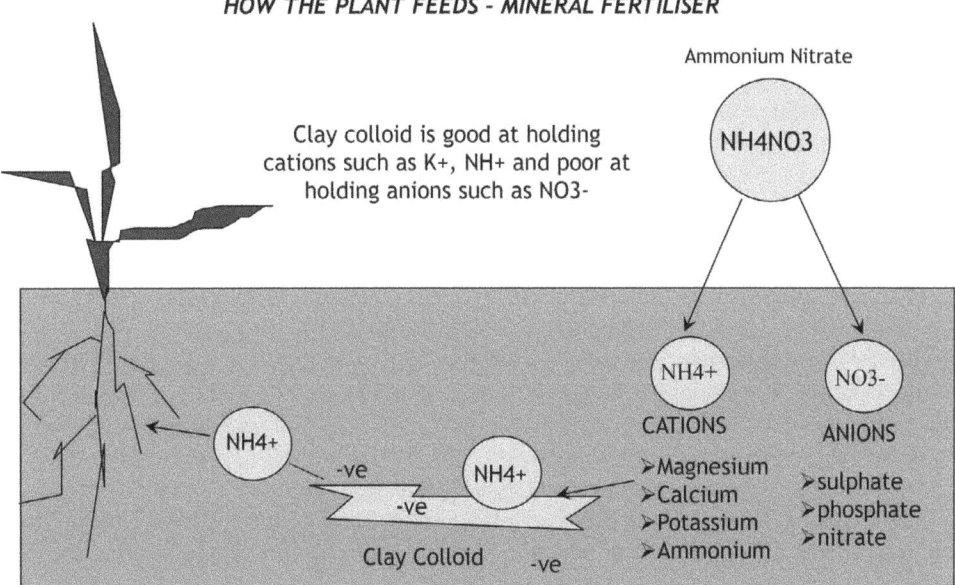

HOW THE PLANT FEEDS - MINERAL FERTILISER

Ammonium Nitrate

Clay colloid is good at holding cations such as K+, NH+ and poor at holding anions such as NO3-

NH4NO3

NH4+

NH4+

-ve

NH4+

-ve

Clay Colloid   -ve

NH4+   NO3-

CATIONS   ANIONS

➤Magnesium
➤Calcium   ➤sulphate
➤Potassium   ➤phosphate
➤Ammonium   ➤nitrate

When mineral fertilisers such as Ammonium nitrate are applied, the cations are held in the soil colloid "bank" which also holds water. However, rain will take nearly half of the nitrate into groundwater.

Fig 12.1: How the Plant Feeds – Mineral Fertiliser

This system does occur, as depicted, in hydroponic crop production where nutrients are in solution. Although not incorrect in soil-based crop production, especially when referring to agriculture which uses mineral fertiliser, this description is incomplete and potentially misleading, especially if the concept of this loop were applied to soils with significant levels of humus or organic matter. Further, it does not explain why natural ecosystems don't leak enough to cause pollution. Figure 12.2 shows how such pollution is avoided and shows the mycorrhizal conduit which is the central mechanism in what is commonly referred to as the *closed loop*. It is the mechanism which stops leakage at a level of pollution. It is this very same mechanism which feeds plants and protects them from disease.[81]

In natural ecosystems, plant nutrients do not enter into solution in the ground water in order to enter the plant. Humus is a complex mixture of heavy molecules which are not soluble in water. Neal Kinsey, in his book *Hands-on Agronomy*[82] points out that this humus has several times the colloidal capacity of clay and will hold onto anions as well as cations. That, however, still did not explain how the nutrients got into the plant without leakage. It was the American PGA (Professional Golfers Association), back in the1980/90s, who pursued this investigation to show that the soil fungi, known as mycorrhiza, fed at one end of their hyphae on the humus and the other end went not up to somewhere near the plant root hair but actually *crossed the root hair wall into the plant.*

81.   Butterworth, B: *How to Make On-Farm Composting Work*, *MX Publishing*, London, 2009

82.   Kinsey, N: *Hands on Agronomy*, *Acres USA*, 1999

This finding was added to by researchers at Aberystwyth in South Wales who showed that there was another type of mycorrhiza which went up to the root hair and wrapped around it much as the placenta in a mammal. This is a molecular level relationship and a closed conduit. That is why the natural ecosystems do not leak.

What composting can do is provide a 'buffer' between a controlled process and the soil. That buffer can isolate physical, chemical and biological risks in order to allow processing, monitoring and safety controls to operate.

Why Soil Universes Do Not Pollute Themselves

Soils manage pollution in two ways. It is, perhaps, useful in understanding how these mechanisms work to first understand what pollution actually is. Pollution is, incidentally, not just quite natural but fundamental to life itself. *Part of* the definition of a living organism (as distinct from not living – a tractor or a computer for example) is that the living organism is producing pollutants. These pollutants are products from the body of the organism which must be got rid of, outside that body. The question then arises as to when that production of pollutants becomes 'pollution'. The answer, in the scientific sense, is *always*. In the practical or legal sense, the exact definition of pollution depends on the following. If the production of pollutants is at a level where the local environment cannot, in time, bring the system back to what the ecosystem previously was, if there is a shift in the ecological equilibrium, then, it may be said, pollution may have occurred. At the end of any discussion, pollution is in the eye of the regulator which tends to restrict the innovation we need.

The two ways a soil combats pollution are by providing a 'buffer' to buy time and by digesting the pollutants and passing them into the ecological chain. Biological systems are rarely instant, and commonly not even fast. Certainly, digestion is slow and the bigger the molecule, the slower the process. However, given time and possibly enough dilution, nature will digest almost anything.

Soils which are substantially sands have little buffering capacity and little ability to hold crop nutrients. For example, by adding ammonium nitrate fertiliser to sand, the ammonium cations and the nitrate anions will leach out very easily with probably more than half going into the groundwater with rain or irrigation. That is a significant economic loss and potential pollution of groundwater. However, add the same material to a clay soil and the colloidal capacity of the clay will retain much of the ammonium cations and possibly some of the nitrate anions, too. Add the same material to humus and there will be a retention of most, maybe all, of both ions.[83,84] There will be no leaching with rain or irrigation of either the ammonium or the nitrate ion and pollution of groundwater will be eliminated[85,86] So, different soils will have different buffering effects and we can alter that capacity by adding and managing the organic matter levels of soils, specifically the humus content.

83. Butterworth, B: *Clamping Down on Compost, Resource, American Society of Agricultural and Biological Engineers*, April 2006

84. Butterworth, B: *How to Make On-Farm Composting Work, MX Publishing*, London, 2009

85. Kinsey, N: *Hands on Agronomy, Acres USA*, 1999

86. Rose, SC et al: *The Design of a Pesticide and Washdown Facility, British Crop Protection Council Symposium*, November 2001

HOW THE PLANT FEEDS – NATURAL ECO-SYSTEM

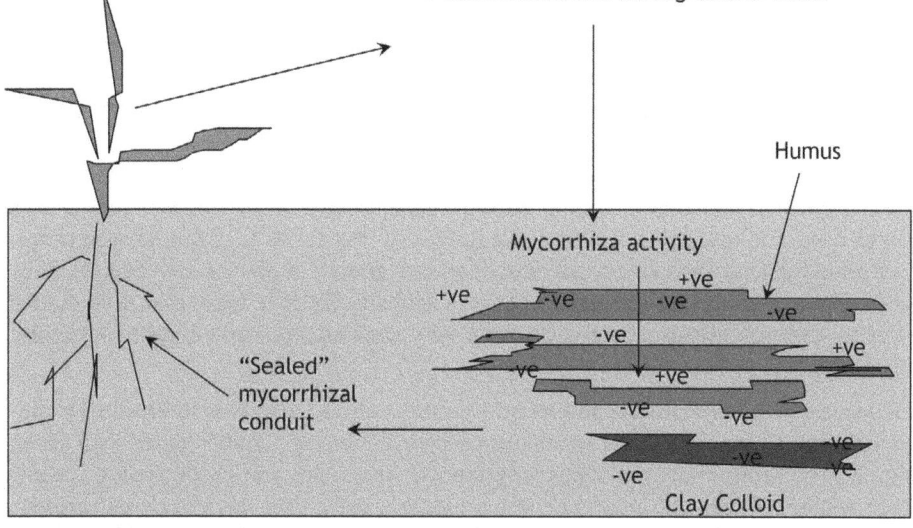

Mycorrhiza are the key to pollution control because they give a "Closed Loop" to recycling both cations <u>and</u> anions. Organic-based systems do not leak nutrients.

Fig 12.2: How the Plant Feeds – Natural Ecosystem

HOW THE PLANT FEEDS – RECYCLING WASTE

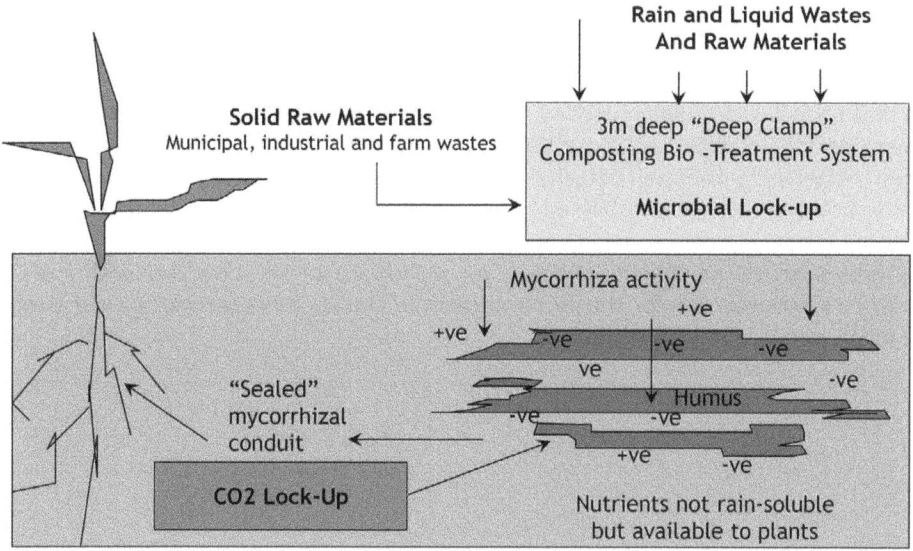

How the closed loop gives pollution control and scope for treatment systems.

Fig 12.3: How the Plant Feeds – Recycling System

An interesting trial was conducted by Rose *et al*[87] in a contract commissioned by Defra on the problems of spills or washings from field crop sprayers. These washings would be expected to contain very active ingredients which, in any concentrated amount released in an uncontrolled way, could cause pollution. The trial involved adding a cocktail of commonly used spray chemical active ingredients onto a series of filter traps such as gravel or sand. One of these traps was half a metre deep with straw, 'non-peat' compost and loam. The last trap, they called it a *bio-bed*, was almost totally effective in breaking down the substances and eliminating pollution risk. That bio-bed was half a metre deep and worked at ambient temperatures which, in the UK, would rarely be above twenty to 25°C. In *deep clamp* composting, Land Research Ltd[88] uses a bio-bed which is three metres deep and maintains temperature ranges of fifty to 90°C, giving several times the absorptive capacity (less risk of effluent seepage), greater buffering effect (gives time for the process to work) and greater process capability. So, it is clear that soils, and even more so, compost heaps, can and do have very sophisticated and capable antipollution capabilities.

This leads us to a conclusion and, of course, naturally, to a further question. Firstly, yes, it is possible to manage these mechanisms by introducing a buffer (such as a compost heap in a controlled and managed, 'imprisoned' situation) and by providing conditions which will encourage and manage the biological activity, thus destroying pathogens and undesirable toxic molecules. The question is how to identify and manage the need for sufficient time and dilution to allow these desirable functions to reach an identifiable and acceptable end point within a predictable time frame. The answer to this is known as *dispersion technology*.

There is a practical, environmental difference between a point risk and a dispersed risk. When a risk is identified, such as a possible pollutant in a compost heap, then the natural instinct of many, including the regulators, is to contain that risk at that site and seek to process it out. That, indeed, maybe the best solution. Alternatively, it may be a lower risk to spread the pollutant to agricultural land in order to disperse it. On the face of it, that sounds like an uncontrolled risk compared with a concentrated, point risk in a compost heap. However, suppose the concentrated risk at the compost site leaks out – then there maybe pollution that the local environment may take some time to deal with. Alternatively, if the risk is dispersed by spreading to land which is biologically active, then the soil micro-organisms may deal with it. For dispersion to work successfully, the soil must be adequately bio-active and dilution must be to a level it can deal with, and there must be a balanced diet for the micro-organisms. Detail of this technology is the subject of much of the following chapters.

87.   Ibid.

88.   Butterworth, B: *Survival – Sustainable Energy, Wastes, Shale Gas, and The Land*, *Amazon*, 2016

Chapter 13: Soil Carbon, Bio-Active Soils & BACS

Building the Bio-Active Carbon Reservoir or 'Sump'

So, building a reservoir of organic carbon will remove carbon dioxide from the atmosphere and improve soil fertility and enable more food production per ha. However, as these pages will show (see *Chapter 19*) soil organic matter will oxidise away, especially if near the surface and mechanically cultivated. 'Near the surface' will vary according to soil type and conditions. For example, with a dry sand rapid oxidation will occur probably down to 100 or 150 mm (four to five inches). In a clay, this will be reduced to maybe fifty mm. Therefore, logically, the best thing to do with compost is to plough it in, preferably below 150 mm. How much? There is no limit. There is no risk to groundwater from nitrogen leaching if 10,000 tonnes per ha, or more, were ploughed in. How do we know this? Look at the Fens in the east of the UK. When Vermuyden drained them some 300 years ago, some of the black organic soils left there have grown crops, the best in the country, with exported harvest, without ever adding fertiliser of any sort. Clearly, the nutrient reserves, especially nitrogen, were enormous, yet the dykes and Norfolk Broads are not polluted.

The figure below shows the basic construction of a sub-surface reservoir of organic carbon. Compost of any quantity can be put into the reservoir, mixed with enough of the local soil to create enough soil strength to support the field traffic (machinery) used in that situation. The proportion will vary according to the local soil and the constituents in the compost, particularly particle size but, as a guide, a 50/50 mix.

Fig 13.2 over the page shows an option for low grade compost or compost contaminated with trash such as plastic bottles. This can be used as hedgerow boundaries or forestry plantings. In hot climates (see *Chapter 20*) this can be used for planting energy crops such as jatropha. The purpose of the polypropylene netting is to stop wind-blow.

Sub-surface organic Carbon reservoir

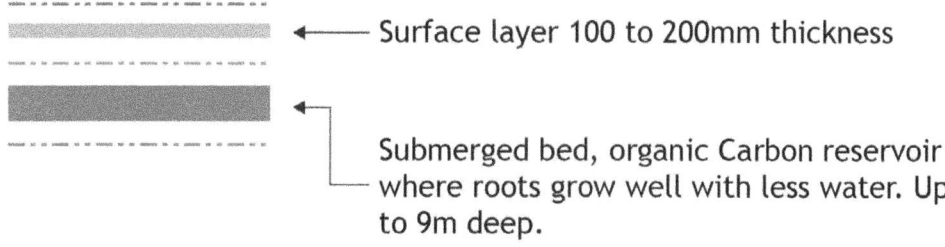

← Surface layer 100 to 200mm thickness

Submerged bed, organic Carbon reservoir where roots grow well with less water. Up to 9m deep.

Fig 13.1: Sub-Surface Carbon Reservoirs

TOP SOIL RESERVOIRS

TRENCH RESERVOIRS

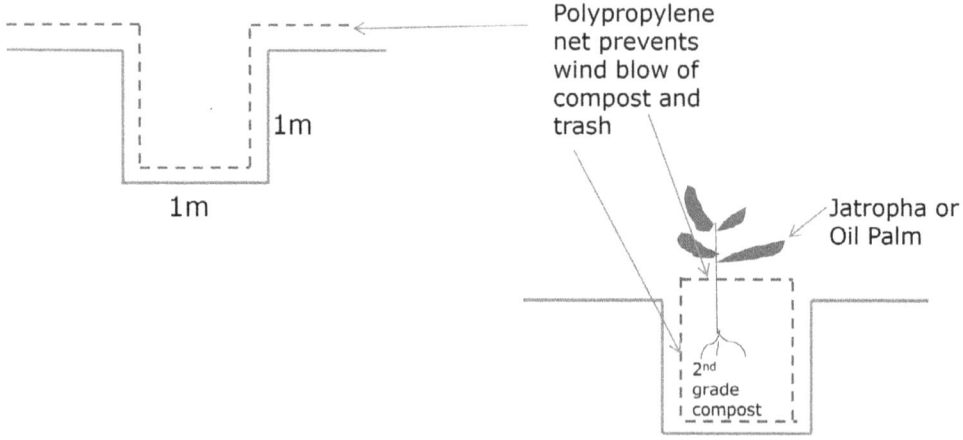

Fig 13.2:Trench Reservoirs

Chapter 14: Using Wastes to Feed Photosynthesis

Composting, Feeding or Direct Incorporation into Soils

More than 99% of all scientists are still alive. Technological change has just reached the steep part of the exponential curve. We already have new technologies and we will have more. Some will be suitable for siting on farms, some not.

When this development of recycle to land was first conceived under the original DTI programme, there was a need to choose a method of operation. The textbook systems were examined, especially the then fashionable windrow system with straddle type windrow turner. Most of the methods discussed in the technical media were of imported technology, using major pieces of equipment and with significant advertising budgets. One by one, these were discarded, not because they did not work (they all did – at least in some sort of way), but because they did not fit the criteria of using what already existed in the rural economy and, most importantly, did not fit the skills and culture that managed the landbank, i.e. farmers.[89]

If differentiating by detail, there are almost as many methods of composting as there are operators. However, there are two fundamentally different approaches – and, of course, every variation in-between.

Windrow composting, strictly speaking, is composting in long thin rows of material. They vary significantly but may be as small in cross section as one metre deep and three metres wide, turned with a multi-tined rotor machine that straddles the row. The other distinct type is known as static pile, table-top windrows or *deep clamp composting* (DCC). The last one is a large heap, around three to four metres deep and is turned with some sort of mechanical shovel or 360° excavator bucket.

At the time of having to make a choice, back in the early 1990s, it was necessary to look at the classic windrow technique using a straddle-type turning machine. Even then, the price of such a machine would have been over £100,000 (GBP equivalent to, say, US$150,000). The windrows would have been, typically, maybe one metre deep in the centre and three metres wide. When it rained, the windrow would thatch in heavy rain and the run-off would run in a stream down each side. There would be a question of mixing the edges to the central heated area. Of major concern was, and still is, the fact that the rotor was designed to mix air with all of the windrow. The centre of that windrow would contain bacteria (some of which might still be pathogenic in the early stages of composting), fungi and volatile chemicals. These would be raised into the surrounding atmosphere. Indeed, when one of these machines is working, there is a cloud of steam around the machine. That steam would be laden with those organisms listed above and now often referred to as *bioaerosols*. There was a perception of health risk. But what really directed a look elsewhere was cost and failure to use what already existed on the ground.

---

89. Empirical experience in the *Land Network* consortium of farmers who recycle wastes to land. Some of this experience is published, much is held by Land Research Ltd.

For those of us at the time who had grown up in the agricultural industry, it was common knowledge that farmers had, for centuries, been making large heaps of 'wastes' in a corner of their fields. Indeed, it was one of only two ways of making fertiliser for the land. One was to use green manure crops including legumes which could fix their own nitrogen. The second was to use 'wastes'. Most people, even of urban background, knew that farmers used the manure from livestock to fertilise land. If the stock had been fed on food imported onto the farm, such as grain (as opposed to home grown food) then the nutrients available to put to land would increase. What few people in the UK can now remember is that farmers also routinely took local wastes and put them to land usually via a large storage heap. The storage heap would be slowly built up and it would stay there maybe two or three years, sometimes much longer, and wait for rotting down and a suitable crop window. They did not usually call it 'composting' but that is what it was. Interestingly, they usually did not turn it at all. The material was not shredded either. It was a slow process and it did not usually smell. Surprising to many today, this was a common, normal practice, without odour, nuisance or pollution.

At the time, as now, the choice of the design of the heap was very important. The shape was clearly to be flat-topped to absorb the rain, necessary as moisture in the process and to stop run-off. But what depth should the heap be? At the time, it was generally acccepted that a small garden compost heap needed to be around 1.5 metres deep to avoid rain running through. So, arbitrarily, that figure was doubled to three metres. Later it emerged that this allowed the best compromise of letting the heat out and the air in.

HOW DEEP CLAMP WORKS

OBJECTIVES
1. To produce stable humus
2. To aerate the clamp, maintain moisture and retain heat up to 55°C
2. "Boil off" surplus water
3. To thoroughly pasteurise the products
4. Kill weed seeds

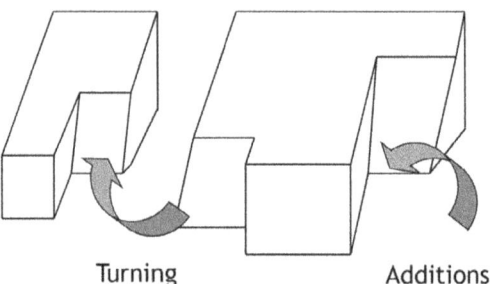

Turning   Additions

RISK CONTROLS
1. Start with straw or green wastes to aerate mixture of waste
2. Make clamp 3m deep from municipal, industrial and farm wastes
3. Manage turning to let air in and control temperature in the 55 to 80°C range

Fig 14.1: How Deep Clamp Composting Works

The table below compares two composting methods. On the left, windrow composting. On the right, a large heap with enough depth to absorb rainfall and hold it. If the heap were vertical sided and the rain vertical, then the rain outside the heap had fallen there for millions of years without harm to the environment. In practice, the sides of the heap are not vertical, nor, depending on the wind, is the rain; so there is a small edge effect which the soil can easily deal with.

So, the research and thinking at the time moved away from the windrow system. This was partly because, at the time and in farming, time in the process was not important. The research conclusions were summarised at the time as in the following table.

Table of Comparison: *Windrow Composting vs Deep Clamp*
Source: DTI research contracts carried out by Bill Butterworth between 1992 and 1996.[90]

This research programme looked at the deficiencies of centralised composting in general and windrowing in particular. The conclusion was to develop existing garden and farm composting techniques and work on de-centralised, rurally-based systems which recognised:

- where the market was; i.e. UK farming then bought around £900m (GBP equivalent to around US$1.5 billion) worth of fertilisers pa and needed the organic matter;
- where the resources already existed space, machinery, labour, skills; and
- farm-based systems are mainly responsible, but centralised systems with pressurised, target-meeting, absentee managers and go-home-at-5pm operators are not.

Windrow	Deep Clamp
Operating Area Commercial input example of 1000t would take approaching 1ha to set up. This is normally on concrete.	As little as 25% even down to 10% of the area. Often not on concrete, just hardcore (since that original choice, the UK Environment Agency now insists on concrete).
Rainfall & Run-Off In heavy rainfall, for example above, there would be 2400 meters of dirty brown water running down the sides of the windrows.	In a well managed clamp, total elimination of run-off and leachate. The operation actually gets short of water and may need to import liquids for best results.
Pasteurisation Poor surface to volume ratio with approximately eight sq m of surface area per tonne and shallow depth of insulation. Therefore poor heat retention, especially at the edges and poor pasteurisation.	About 0.18 sq m of surface area per tonne and 3m depth. Easy to control and continuous good pasteurisation. At least 6-log reduction in pathogens (one millionth) and maybe seven log (one ten millionth).
Chipping Necessary either by pre-chipping or by the rotary windrow turner.	Not necessary but some sort of shredding is normal to speed up the process.

90.   Ibid.

Windrow	Deep Clamp
Process time Financially important. Usually four to twelve weeks.	Financially not important. Not usually less than windrowing. If thick material not chipped, then much longer.
Odour Probably low odour but some materials difficult.	Less odour. Maybe very little odour with most materials. Can manage well to give low odour on very difficult materials.
Mechanisation Normally requires large, expensive and not-road-mobile machinery.	Normally operated with existing farm machinery.
Centralised Because of the machinery and concrete, normally centralised. Therefore major operation and probable NIMBY ('not in my back yard') opposition.	Normally de-centralised with 65% to 85% less transport distances. Small scale farm operation with a degree of isolation. NIMBY very rare.
End Product Normally needs shipment elsewhere.	Normally used on site. Possibly a contribution of up to £900 million to UK balance of payments.
Production of Aerosols & Airborne Fungal Spores A major problem at turning. Probable risk to health. Future regulation may necessitate building cover and air filtration.	Possibly a 7-log reduction (i.e. one ten millionth).

Shredding and Screening

The problem with mechanical treatment such as shredding is that it requires a lot of energy and, therefore, has a cost both financially and environmentally. Big shredders may consume fifty litres of diesel fuel per hour, or more. The advantage is that smaller particles will be easier to handle mechanically and the process will be faster. However, green waste from twin-bin municipal collections very often will not really need shredding; particularly if the final product is to be put to heavy soils where large residual particle size is an advantage. So, the garden compost as a fine, uniform, black 'crumb' is not what is required in most arable farming operations. This implies that screening, to remove the larger bits, may not be needed. Indeed, most of the farms in the national farmers' consortium, *Land Network*, so far, do not screen at all, or only selectively. Screening, however, may be an alternative to shredding. For example, with suitable twin bin green wastes (which may be of low branch size but are often heavily contaminated with plastic and other household wastes) it may be better on all counts to put the whole lot into the composting process without shredding. Screening out at the end will make the plastic much easier to pick and the oversize particles can just go round again.

Outline of the Basic Process in Composting

i) Materials are delivered to a farm site, with or without concrete, with no building of any sort, and stacked in a heap three metres deep. The reason for no concrete is that concrete collects rain and allows it to drain into the *bottom* of the heap. Rain going in at the *top* of the heap will steam off because of the heat generated in the heap. The heap is fairly tidy with sides usually at around 45°C. This three-metre depth was originally based on the gut feel of experience. There has been some significant research[91] on this potentially critical aspect of compost process temperature since. This confirmed the functional desirability of a deep heap in the on-farm situation. A deep heap can and will remain aerobic provided the shredding of the material is relatively coarse. There is a further characteristic of the deep heap and one which initially appears to be a disadvantage as it often means the process is not quite as fast as windrow composting and is often a little cooled. The advantage was shown later with research on best anti-biotic activity, carried out at the University of Hull, which showed that 55 to 60°C, for a longer period than processing at 60°C plus, gave better pathogen kill.

ii) In the UK, this recycling of waste to land could, back in the early 1990s, only legally be operated with an exemption from the *environmental permitting regulations* registered with the Environment Agency, or with a full scale, much more involved, full licence or permit. Again at that original time of writing, for the exemption, amongst other restrictions, this entailed operations with not more than 1,000cu m in the process at any one time. This, at three metres deep, will be approximately eighteen metres square, i.e. a little less than a cricket pitch down each side. This original regulatory climate allowed a farm to start small scale and progress upwards, one step at a time. This UK regulation has become progressively restricting. By the time of writing (2019) composting above fifty tonnes pa needed a full permit, involving concrete. This progression has meant that it is very difficult for farmers not already composting to start up. This is a fundamental failure of government to discipline the regulator to enable progress. Whatever the regulations, there is no technical limit to the size of the operation.

iii) There is some significant technology in the rules of how the heap is put together and managed so as to prevent leachate and prevent anaerobic conditions. This can be controlled in practice with a relevant code of practice and licensing of operators by the regulator (see *Chapter 15*).

iv) 100% of the material in the *Land Network* programme is spread to land and incorporated (ploughed or cultivated to mix in with the field soil) when there is a crop window. It may also be spread behind a silage cut. Finely screened composts can be put to growing crops, including grass. The material is almost completely odourless on maturity and spreading.

91. Ibid.

Managing Odour Through the Temperature Curve

Look again at the basic temperature curve in the compost process, repeated below. The basic temperature curve is never, in practice, as tidy as is shown in the graph. The main risk of really bad odour is during the first peak – the bacterial phase. High activity and low oxygen supply mean the process is likely to be anaerobic and bacterial and offensively odorous. Going anaerobic in the trough (when actinomycetes take over) is not likely to have serious risk of bad odour and going anaerobic in the second peak and onwards will certainly make any 'musty' smell of the fungi worse – but unlikely to be as offensive to the human nose as anaerobic activity in the first, bacterial phase.

COMPOST TEMPERATURE CURVE

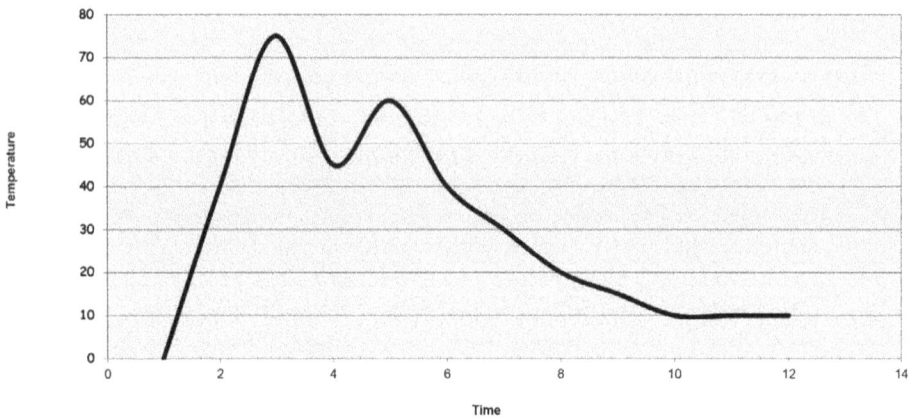

FOUR BASIC PROGRESSORS
1. Live Organisms
2. Food
3. Moisture
4. Oxygen
If temperature is off-curve, one of these four is sub-optimal

Fig 14.2: The Basic Temperature Curve

In reality, some materials will process faster than others. The more rapid the process, the more likely it is to be relatively uniform and at a particular stage. Therefore, in very rapid situations, such as with catering wastes and with rotary turning every couple of days, the risk of odour in the first stage is at its highest. Conversely, in slow situations such as with woody green wastes, not turned very much in *deep clamp* processing; only some of the material will be active, and probably not very active, and, therefore, the risk of odour is the lowest.

The simplistic conclusion on odour control is that the highest risk is in the first peak phase; anaerobic bacterial action here is likely to be offensive. There will always be some mustiness in the second peak and onwards and this will be made worse by lack of turning if the mass of the compost is active. There is a way of reducing the most offensive risk. Consider the following variation.

NORMAL AND LOW TEMPERATURE PROCESS CURVES

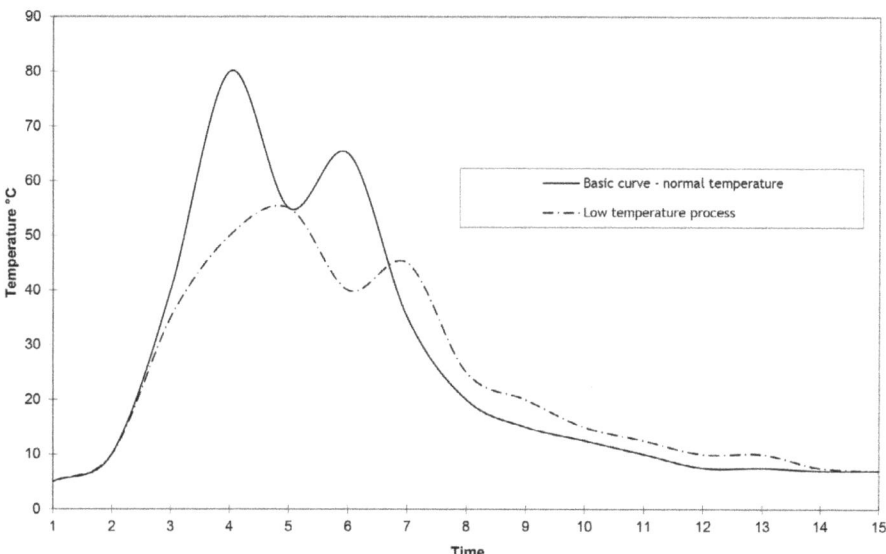

Fig 14.3: Normal and Low Temperature Process Curves

The second curve above shows a lower first peak temperature. If temperature can be limited to less than around 65°C, for example by not having quite enough of a particular food which the micro-organisms need (commonly, this is nitrogen) and by not reducing particle size too far, then the rate of micro-organism activity will reduce and the risk of odour in this phase will also reduce. This can be easily managed with deep clamp, or 'static pile' composting; it is slower, but safer.

In practice, one of the advantages of DCC is that it can be slower and because it has a very low surface to volume ratio, the scope for odour is also very low. If it is run with large particle size (so that it can breathe), of not very high nitrogen materials, and with little turning in the first phase, then there is not likely to be significant odour of the offensive, anaerobic bacteria type which can, sometimes, emit from that first period of the process. However, the rest of the process, as always, is likely to have a musty odour, characteristic of fungi. Again, running a slow process, with little turning, will minimise emissions. In practice, there is a compromise between turning to keep the process aerobic, slowing it down by not turning to reduce emissions, and the speed which is required for commercial or regulatory reasons.

Aeration by Suction from Below and Bio-Beds

Where odour is an actual or potential risk, by far the most effective, and least cost, solution is low volume suction from below the heap, feeding into a bio-bed.

If air is blown through a granular or mixed particle material, it will tend to 'track' through the pathway of least resistance. This leaves some areas over-ventilated and some under-ventilated. What happens here is that air in the voids travels at a rate which becomes turbulent. At what is known as the critical flow rate, that turbulence becomes so violent that the air just goes round and round, rather than travel to the next void. Almost always, putting more power into the fan and increasing air flow, increases this turbulence effect, making the lack of uniform ventilation worse, not better.

Sucking air from the bottom of the heap at very low volumes, avoids this turbulence and results in a uniform air flow through all the voids. A fan of only one to two kW will ventilate several hundred tonnes of compost.

If the extracted air is pumped (with the same fan) through a bio-bed which will extract all the odour once it has become biologically active. The bed is made from coarsely shredded wood, preferably root timber which does not degrade (compost) so quickly. Alternatively, the bed can be made of plastic particles but what is needed is a large surface area. Natural wood, especially root timber, will already be 'dirty' and carry micro-organisms and it is these that will digest the odour and anything else in the airflow. The bed will be around one m deep and take a couple of weeks to become biologically active. It will take many months but, eventually, the process will digest the wood chip and, as the particles become smaller, the resistance to the airflow will increase, and the chip will need replacing.

Managing the Crop Nutrients in Wastes

All materials, not just wastes, contain, at least to some extent, 'food' for micro-organisms. Natural ecosystems are insidiously capable of reconstructing almost everything through closed loop processing. Even the Titanic, sunk two miles down, is not rusting in the absence of enough oxygen, but it is being eaten by bacteria. It may take several hundred years, but they will do it. This demonstrates something of regulatory significance. Very often, those not skilled in the art, with limited technical knowledge and/or experience, will say something won't work or is impossible. Most things work in nature; it will handle anything given time. The question is more about how long it will take and how to manage and vary that time to fit the farming seasonal work load.

There is a word of caution in thinking about carbon and its use in soils. Carbon certainly can be involved in large molecules which are, in the language of an organic chemist, 'organic'. Carbon may also have a very significant effect on the fertility soils when it is present as pure carbon, in its amorphous or activated state, as shown in the *terra preta* soils of the South Americas.[92,93,94]

Micro-organisms certainly need food and they need moisture and gas exchange – if we are interested (as we are) in aerobic composting, they need oxygen from the air and need to be able to breathe out carbon dioxide, just as other forms of life, including humans. But what do we mean by 'food' and how flexible are they? Bearing in mind that farmers in the UK have now been ploughing in six to eight million tonnes of straw per

92. Mann, CC: *Our Good Earth*, *National Geographic*, September 2008, pp80–107

93. Sombroek, WG: *Amazon Soils*, 1966

94. Butterworth, B: *The Straw Manual, Spon*, 1986

annum for over forty years in the UK[95] does a text-book view of 'carbon : nitrogen ratio' really matter?

The 'food' that micro-organisms need is basically carbon (for energy – just like sugar and carbohydrates are to us) plus all the other nutrients that our own bodies live on. The carbon can be in the form of almost any organic molecule (i.e. any molecule containing carbon linked to other carbon atoms, hydrogen and possibly many other elements). They also need a balance of potassium, phosphate, calcium, sulphur, magnesium and the full range of trace elements, just as we do. Micro-organisms do differ slightly to humans though in their ability to use a wide range of nitrogen sources to build their own body proteins. Unlike humans, they can use non-organic forms including, in the case of the bacteria attached to the roots of legumes, atmospheric nitrogen. They also have the flexibility to be able to adjust to wide variations in the concentration of organic carbon molecules. There are, of course, practical limits to how long it will take these organisms to digest hard plastics.

In discussions about C:N ratio, a range is often given, commonly twenty/thirty to one, without reference to whether this refers to the start of the compost process or where the farm wants to end up with in the finished compost. There are many text books, academic papers and standards relating the proportion of carbon to nitrogen in a compost process and product[96]

Basically, micro-organisms need a lot of carbon (for energy) and not so much nitrogen (to make cell body proteins). Many 'authoritative' guides will give, as a guide, an ideal of twenty-five of carbon (by weight) to one of nitrogen. Alternatively, a range of twenty to thirty to one is usually a better guide to what actually works. The truth is that it can be much wider than even that and this is because the micro-organisms will use up some carbon in the process. Furthermore, the more difficult the process, then the longer it will take and more energy, and therefore carbon, they will use up.

As a guide, the following gives an indication of the C:N ratio of a number of materials which may be 'on offer' to a composting operation:

Carbon to Nitrogen Ratio in Materials

Material	C:N Ratio
The best top soil	10:1
Humus	10:1
Kitchen food waste	15:1
Vegetable wastes (factory)	20:1
Lawn mowings (UK May)	10:1
Lawn mowings (UK August)	25:1
Tree leaves (autumn brown)	35:1

95. *Understanding Rural Land Use, Environment Agency*, NE-1/100-3.SK-C-BEKC

96. Butterworth, B: *Busting the Myths, CIWM Journal*, June 2015, pp28–30

Manure (stable- high straw)	50 to 70:1
Manure (cattle, low straw)	20:1
Cereal straw	70 to 100:1
Sawdust	150 to 800:1
Newspapers	150 to 200:1

As an example of handling high carbon inputs successfully, in the mid 1980s British farmers faced a ban on burning cereal straw behind the combine harvester. The question was: what would happen when around 6.5 million tonnes of straw was ploughed in every year? Put another way; how could this be best managed?[97] By the mid 1990s, every farmer was doing it without a second thought and yet the straw had very little nitrogen in it and the C:N ratio was well outside the figures discussed above; straw will, depending on variety and season, have a C:N ratio into the 70–100:1 range.

Digestion of Organic Carbon Wastes

Generally speaking, decomposition by micro-organisms is much the same process as digestion by ourselves in that it requires energy. As we metabolise sugar during exercise, they will metabolise carbon-based molecules to produce the energy to drive the process and produce carbon dioxide. They need nitrogen to build cell proteins but they need much more carbon than nitrogen. In processing the nitrogen, they will metabolise carbon and the more difficult the digestion, the more carbon they will use up. As the C:N ratio rises above 30:1, the process gets more difficult and uses up more carbon and takes longer.

There is one more factor to complicate the issue and the judgement on how well the process will go and whether more nitrogen will be needed. With ploughing in straw, for example, some of the carbon is in the form of cellulose and will quickly break down if there is enough nitrogen in the system. Further, there will be a lot of carbon present as more complicated molecules with more stable cross linkages in the carbon chains, such as hemicelluloses and lignins. These will take longer to process and may not require more nitrogen for the micro-organisms in the short run.

Research, up to the point when straw burning in the field was banned in the early 1980s in the UK, showed clearly that ploughing in too much carbon, too soon, would produce what farmers called *nitrogen starvation*. This meant that the soil micro-organisms would take nitrogen out of the surrounding soil and use it for processing the excess carbon, thus robbing the soil of nitrogen for the current crop which would, then, suffer. However, the soil usually recovered after two or three years and the rules, then, for incorporating straw involved[98] adding some easily available nitrogen supply after the first year of straw incorporation, a little less in the second year and maybe some more in the third. Then the soil mycorrhiza would cope and the soil would support crop yield. These rules would be modified with less nitrogen added if the soil had significant biological activity (because of previous organic matter additions encouraging the populations of mycorrhiza) or

---

97.   Butterworth, B: *The Straw Manual, Spon*, 1986

98.   *Understanding Rural Land Use, Environment Agency*, NE-1/100-3.SK-C-BEKC

increased and lengthened on arable soils depleted of organic matter (and, therefore, the activity of mycorrhiza).

Interestingly, if the supply of nitrogen is less than the C:N ratio of 20:1 (i.e. the material is nitrogen rich) then the soil will increasingly ammonify the nitrogen compounds. If the soil is wet, this may then be washed into the groundwater, in dry, warm conditions, the ammonia may come off as a gas to atmosphere. Thus, another reason, in a composting operation, not to let the heap dry out. With adequate moisture, and preferably with the ultimate addition of enough carbon, the micro-organisms will find a way to convert that nitrogen into large molecules which become part of the humus and, therefore, not leachable and not likely to be released to atmosphere.

Different C:N ratios will produce different compost temperature curves and different processing times. In part, this can be managed by altering the turning regime, but only in part.

Generally, low carbon/high nitrogen materials will produce higher temperatures and process faster, while high carbon/low nitrogen materials will be slower and reach lower temperatures.

Effect of Carbon Nitrogen Ratio on composting process

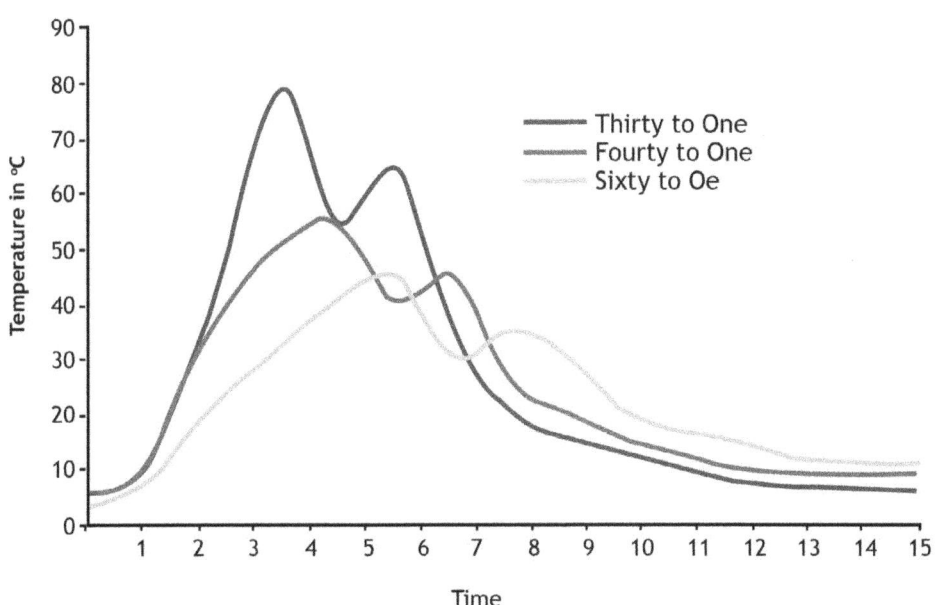

Fig 14.4: The Effect of C:N Ratio on the Compost Process

These rules are, of course, generalisations in a complex environment. For example, if the carbon is present as cellulose, the rate of metabolism by the micro-organisms is potentially much faster than if it is present as hemicelluloses and, again, that too would be faster than if the carbon were present as lignin (i.e. wood). Metabolism of high carbon materials, especially these 'tough' woody materials, will use up carbon. This means that if the feedstock for composting is of very woody material, then the C:N ratio could be high, maybe much higher than sixty or seventy to one and there would be little more detrimental effect then than taking longer to process. This may mean that input materials may be above even 70:1 provided they come down nearer the 30:1 by the end of composting. Despite this, some heavy land farmers are, correctly, happy to see woody particles left in the compost, knowing that they will break down slowly, in that they will help to hold a clay 'open' and let it breathe better, thereby allowing faster soil warming in the spring, better gas exchange with less resultant chlorosis, better moisture movement and, overall, better crop growth. Where the remaining carbon is as lignin, then it will only slowly be metabolized in the soil and is unlikely to significantly affect the supply of nitrogen to a crop. The best advice to farmers,[99,100,101] is to apply the compost in any way possible and, in the early years of converting to compost farming, watch the chlorosis in the crop, especially in early spring, and use instinct to top dress with mineral nitrogen fertiliser if necessary. When well into compost farming, the biological activity in the soils will mean that that precaution is unlikely to be necessary.

In commercial operations where the objective is a low priced material which can legitimately be called 'compost', then it may be that the uninformed will sell composts where the C:N ratio is well above 30:1. Such a compost may be friable and dark in colour but that does not mean that it will support plant growth at all well. These composts may well turn out to be less than reasonable in performance and seedlings planted out into it may fail, quite likely, completely. It is reasonable for the purchaser to expect plants to grow and, therefore, that product could reasonably be 'unfit for purpose' in law. This, of course, is the reason for standards and the current UK standard, PAS100, which is, unfortunately, the only one. It is not particularly suited to farm requirements and it has a very restricted range of permitted materials and, therefore, inhibits recycling to land. However, at least, it is a start.

If the value of the input materials is known, then it is possible to predict the value of the output in terms of value in the soil and to a plant or crop. As a guide, composting micro-organisms will use about thirty parts carbon for each part of nitrogen, and such an initial C:N (available quantity) ratio of 30:1 promotes rapid composting. Researchers report optimum values from twenty to 31:1. A majority of investigators believe that for C:N ratios above 30:1 there will be little loss of nitrogen. University of California studies on materials with an initial C:N ratio varying from twenty to seventy-eight and nitrogen contents varying from 0.52% to 1.74% indicate that initial C:N ratio of thirty to thirty-five was optimum. These reported optimum C:N ratios may include some carbon which

99.   Butterworth, B: *How to Make On-Farm Composting Work*, *MX Publishing*, London, 2009

100. Butterworth, B: *Reversing Global Warming*, *Refocus*, September/October 2006

101. Berg, B et al: *Plant Litter, Decomposition, Humus Formation, Carbon Sequestration, Springer*, 2003. This is one example of many hundreds of references to organic matter degradation and formation of humus in the academic literature and on the web

was present as hemicelluloses or lignin. Composting time increases with the C:N ratio above thirty to forty. If proportion of large molecule carbon is small, the C:N ratio can be reduced by bacteria to as low a value as 10:1. Values of 14:1 to 20:1 are common depending upon the original material from which the humus was formed. These studies also showed that composting a material with a higher C:N ratio would not be harmful to the soil, however, because the remaining carbon is so slowly available that nitrogen starvation would not be significant.

Does all of this really matter? Well, the C:N ratio will certainly affect the rate of composting process and how plants will grow if planted into pure, undiluted compost. However, seeds and plants in agriculture are not planted into pure compost. In this case, it is how the addition of compost will affect the soil as a whole and into which a crop will be sown and grown. If the technical base is assumed to be purely chemical, then it may be that an uncomplicated view of the chemical value of the compost is a sufficient guide. However, the technology indicates that there is a wider biological effect and the carbon input into an arable soil with run-down organic matter may well benefit from higher inputs of carbon because the soil mycorrhiza will use up carbon in their energy systems, just as is the case with the composting organisms. Frankly, in farming, the soil will cope with quite wide values, especially after a few years of applications because the biological activity will buffer the extremes and still feed the crop with what it needs.

There is one area of use of large carbon chain molecules in composting which has been sadly neglected in developed agriculture and with results which have sometimes been visibly catastrophic but, maybe universally, have allowed a progressive and often un-noticed erosion.[102] Fibres of root hairs, lignin, and synthetic carbon molecules all act as stabilisers of the physical structure of soils. They help to allow carbon dioxide from soil micro-organism activity out of the soil and oxygen back in. They allow water into the soil (reducing surface run off and flood and erosion risk). They help hold water (reducing irrigation need). They reduce cultivation need and energy used in cultivations when they are needed. In terms of soil physics, it does not matter if the carbon molecules are synthetic or 'natural', nor does it matter how long it takes to break them down – in fact, it is better if they just stay there. After all, most people who know anything about soils will accept that peat is a valuable asset. Well, the fibres in peat may be thousands of years old.

Studies of black soils (terra preta) in South America[103,104,105] have put another angle on carbon and its value as amorphous or activated carbon, usually (it is thought) derived from charcoal. This charcoal is sometimes now trendily called biochar. A study at the University of Edinburgh is looking at the potential to use biochar as a means of locking up carbon from the atmosphere in a fairly permanent way and so as to make soils more fertile at the same time.

102. *Every Bottle of Prosecco May Erode 4.4 Kilograms of Italian Hillside*, *New Scientist*, January 21st 2019

103. Berg, B et al: *Plant Litter, Decomposition, Humus Formation, Carbon Sequestration*, *Springer*, 2003. This is one example of many hundreds of references to organic matter degradation and formation of humus in the academic literature and on the web

104. Mann, CC: *Our Good Earth*, *National Geographic*, September 2008, pp80–107

105. Sombroek, WG: *Amazon Soils*, 1966

Terra preta soils show a very high degree of sustainable fertility which, in areas of high temperature and violent fluctuations in soil moisture, might at first appear surprising. Initial studies on *Land Research* farms indicate that the carbon in printing inks appears to be of similar value. The carbon is capable of holding onto nutrients at a level similar to humus. However, humus is easily and rapidly oxidised in such conditions as were found in South America, i.e. of high temperature and moisture, but the pure carbon is not. It may be that there is potential for the wider use of carbon in soils in ways that we do not yet fully understand.

Deep Active Bedding and Outdoor Corrals

It is technically and practically quite attractive to build up a deep pile of organic material and let animals, particularly pigs, do the turning. The plan here is to use a fence to retain the animals, possibly and usefully a wall of large bales of straw, and put a layer of 'waste' on the ground, followed by a period of *grazing* by the stock. Sheep and cattle will do this but keep the layers thin and add new material little and often. The layers can be thicker and of less critical periods between additions by using pigs which will root through the material with their snouts and aerate the mass to maybe up to half a metre deep.

If the 'waste' has significant food value to the animals concerned, it may be possible to avoid adding food supplementation, the animals living entirely on the waste. A variation on this is to run 'barley' beef cattle on the deep bed and run pigs underneath them. The pigs will find enough to eat (including of the cow dung) without supplementation. With care in stock density, this can work quite well.

DEEP ACTIVE BEDDING SYSTEM FOR FEEDING- COMPOSTING OUTSIDE

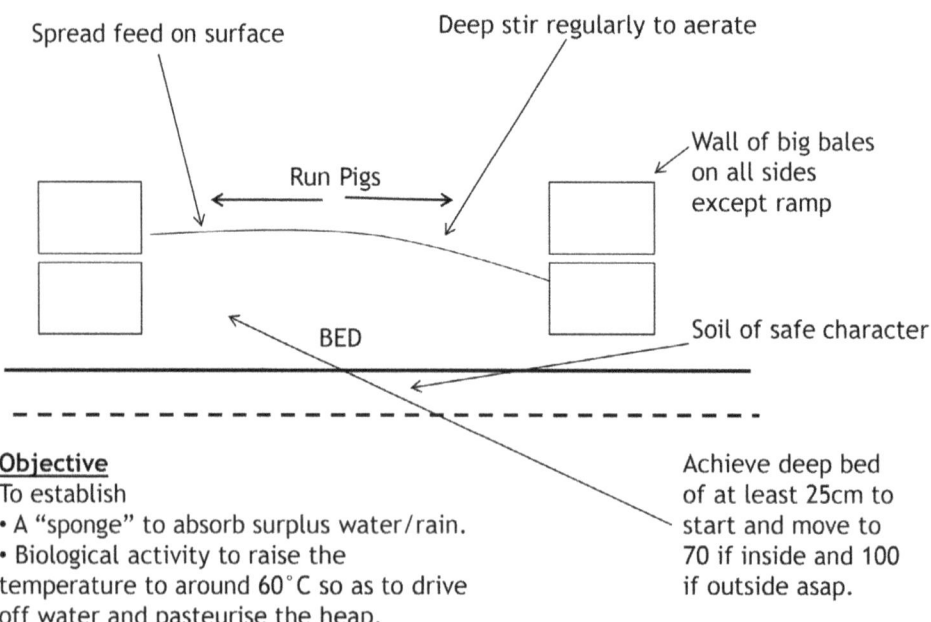

Spread feed on surface

Deep stir regularly to aerate

Run Pigs

Wall of big bales on all sides except ramp

BED

Soil of safe character

Objective
To establish
• A "sponge" to absorb surplus water/rain.
• Biological activity to raise the temperature to around 60°C so as to drive off water and pasteurise the heap.

Achieve deep bed of at least 25cm to start and move to 70 if inside and 100 if outside asap.

Fig 14.5: Deep Active Bedding for Deeding Outside

A further variation is outdoor stock *corrals* which are based on the idea of a high fence around a well-drained base, with layer upon layer of bedding added. The 'bedding', in this case, being wastes with the sole purpose of bedding rather than having some food value. The advantage of this system is that it does not involve the expense of a roofed building. The disadvantage of the system is that, in times of rain or snow, it involves significant added bedding – more than if the stock were housed under cover. That is significant if the bedding has a cost. However, if the bedding is 'waste' and has an attached income, then the corral becomes a processing factory to turn the waste into farmyard manure. This can be an added facility for recycling to land via a logical, safe and profitable route.

Composting as an Industrial Process

Industrial scale operation has been already introduced above. Much of the developed world operate some form of large scale composting and/or digestion systems. Even un-complicated, out-door composting sites of, say 25,000 tonnes capacity per annum, may involve significant areas of concrete at significant financial and environmental cost (concrete has a significant energy cost).

It is quite possible in practice to run a large composting operation, including the addition of liquids, and avoid run-off of any liquid. However, that does take knowledge, experience and discipline. Because humans sometimes break the rules, the regulators' natural reaction is to insist on concrete to cover the point risk (see *Section 6* for an alternative licensing arrangement to police such risks).

Compost can be carried out by a number of variations:

  i)   *In the open, on or off concrete, with aeration by turning.* The least capital cost and can be fitted easily into farm seasons and logistics. The slowest method and may have an odour risk with some feedstocks.

 ii)   *In the open with aeration by blowing high volume air upwards through ducts under the heap.* A step up in costs and high volume air may track unevenly through the heap, a higher odour risk. A significant energy cost for the far. Possibly a noise problem.

iii)   *In the open with aeration by sucking air at low volume through ducts under the heap.* Lower cost than high volume air blowing systems. Air flow does not track. Lower noise levels. Coupled to a bio-bed, any odour risk can be eliminated.

 iv)   *In vessel with solid compost cascading down the vessel and aeration by blowing air upwards.* Low risks, much higher capital cost than other systems. Some energy cost for the fans.

There is no actual limit to the size of any of these processing plants. However, they are all potentially point risks and the bigger the scale of operation, the bigger the risk if anything does go wrong. Secondly, the logistics of feeding the plant and removing the products may reduce the advantage of scale. Bearing in mind that what the objective is of building the bio-active carbon reservoir in the soil, the logic of farm-based, farm scale operation is inescapable.

Digestion of Organic Carbon Wastes

AD – Anaerobic Digestion

All living creatures 'breathe', feed in some way and produce 'waste' products. All micro-organisms are capable of digestion and do either need oxygen (i.e. they are *aerobic*) or work in the absence of oxygen and produce gases (for example, some produce methane and, because they work in the absence of oxygen are *anaerobic*).

Generally speaking, AD, anaerobic digestion, is a process which produces methane and carbon dioxide. This *biogas* is typically half methane and half carbon dioxide and it will burn but do so much better if the carbon dioxide is removed first (by 'scrubbing' with water). The resultant and relatively pure methane can be burned as fuel in an internal combustion engine, to drive a generator, to produce electricity. In so doing, it will destroy most of the organic carbon to produce the methane.

Digestion of Wastes in Mesophilic Processes
Historically, sewage products have caused very significant public nuisance when spread to land. Raw, untreated sewage can no longer be legally spread to land in the UK or most of Europe. Whenever and wherever it is, it stinks. The first line of processing which many sewage companies use is mesophilic digestion in the absence of air (AD). This allows natural (at UK ambient temperatures) micro-organism process to a temperature of up to around 38°C. There is a very significant reduction in odour and, with sewage and many other organic wastes, there is a usable production of methane. Indeed, this is the basis of AD.

Generally, anaerobic digestion works well with liquids up to about 10% dry matter. As dry matter increases, the process becomes increasingly difficult to manage and costs rise (from grinding up solids in order to allow the process to proceed). Historically, AD for high dry matter operations has really not worked.

Mesophilic AD has been used for many decades in China and other Asian countries on a small scale in one-family digesters to produce gas for cooking. Feedstocks are often all family wastes.

Digestion of wastes in Thermophilic AD Processes
These processes, by definition, involve processes going up to higher temperatures, usually around 60°C to get pasteurisation. Gas production is also increased and, therefore, the organic carbon residue is also yet more reduced.

The world is expanding in population and people need food and warm houses. That takes energy and energy security is directly related to national security in the wider, including military, sense. So, for this reason alone, AD to produce electricity is compelling. Also, where subsidies are available to support AD, an individual can make a profit. Further, the technology of AD is getting better. It is as well to remember that methane is potentially an explosive gas.

AD Liquor
Soluble nutrients from AD liquor enter the plant root in a crop. This happens in solution, or hydroponically. It works, very much as ammonium nitrate fertiliser works and British farming has led the world in crop yields for fifty years.

The problem is that forty to fifty percent of nitrogen in solution (probably manly as ammonium and nitrate ions) is lost to groundwater.[106]

With a humus-based eco-system, where the organic carbon has not been removed, nutrients are held in the humus at molecular level. Humus is a very complex and variable mixture of hydrocarbons, carbo-hydrates and proteins formed from the dead bodies of micro-organisms which digested the wastes or other materials supplied to them. Because these molecules are large, they are not soluble in rain or irrigation and are therefore not lost to ground water. It is true that in AD liquor, there will be some organic carbon, but not much as the process efficiency is trying to turn that carbon into methane. What happens in the natural eco-system process is that the nutrients are fed into the crop through the hyphae of the soil fungi (called mycorrhizae) and these feed at one end on the humus and, at the other end, actually cross through the wall of the crop root hair or, another type, wrap round the root hair like the placenta of a mammal. This, then, is a molecular level, no leakage conduit that feeds the crop *on demand*.

There are two things which can be done with waste recycling to land. If wastes are treated aerobically by either composting or TAD (thermophilic aerobic digestion), then they will follow the natural ecosystem and organic route which is much safer for ground water and does not destroy the organic matter. There is no doubt that these routes are sustainable and 'green'. However, if the nation has decided that it will have AD as a producer of electricity, so be it. In this case, if the liquor is to be put on the land, then application of little and often, during the growing season, will reduce losses but significant losses of nutrients, particularly nitrogen, will occur.[107]

On-Farm Logistics
There is one more problem with AD liquor; it is a large volume of, to varying degrees, odorous liquid. 'Available' (in scientific language, *in solution*) crop nutrients are not allowed to be put on the land during a regulatory 'closed period' in winter because of the risk of pollution of ground water. So, the liquid has to be stored and maybe in sealed storage because of the odour. That storage has significant cost and the liquid needs to be spread in the spring when the crop is growing. Bear in mind that most arable land is sown to crops in the autumn and most winters are wet. So, there is a need to take a possibly thirty-five tonne tanker with injection kit to travel over the wet ground and put injection tines through the crop. Not an attractive thing to do and many farmers would not be comfortable with that.

One option is to use a dribble bar – like a crop sprayer but with flexible hoses trailing onto the ground. Not as good as injection at reducing odour but 80% less odour than spraying into the air. Even so, this still leaves the problems of expensive and specialist storage and a high cost of transporting big volumes.

Another option is to compost the liquor with something like straw or green waste. There will be an odour issue but this may be controlled by low volume air extraction from under the heap. A plus is that the 'available' (soluble) nutrients will be locked up to be organic molecules by thorough composting and that will almost eliminate loses by leaching in rain or irrigation.

106. Butterworth, B: *Busting the Myths*, *CIWM Journal*, June 2015, pp28–30

107. Ibid.

What is really needed here is a dried crumb or granule which can be stored in any shed on the floor or in an IBC, transported easily and in a relatively small volume compared with the original liquid and therefore at a fraction of the cost. These granules could probably be spread with existing fertiliser spreaders and at relatively high output. This would also make the product more marketable. The problem, however, is energy cost. There are dryers in development which use low-grade heat from the AD process, topped up with new, efficient energy delivery technology, i.e. microwave possibly coupled to tumble drying, which will, eventually, make drying an economic option where liquids have problems.

So, if we understand how to reduce nutrient loss, how about understanding how to get the cash out of the AD process? The cash comes from turning the organic matter into methane (which can drive a generator or be burned to produce heat) and carbon dioxide (which is waste to atmosphere). The output of the generator can be measured easily and accurately. The trend in the UK is not to generate electricity on site but to put the methane as 'gas to grid', i.e. if the national gas grid is within economic distance, to put the gas into that.

Whatever the destiny of the gas, the efficiency of conversion of the organic carbon in the feedstock to gas, i.e. 'conversion rate' matters. This rate will depend on the design of the AD plant itself, the feedstock and how the operation is managed. The AD plant, if it is single stage, will have a comparatively low capital cost and will be able to take feedstock of low dry matter, say up to eight or nine per cent, and will not get much of that organic matter converted to methane, maybe down to five or six per cent dry matter if it is well managed (a net conversion of two or three percentage units). There are two stage systems which will cost more and get more gas out. Then there are three stage systems which cost more, can take in up to twelve per cent dry matter and get down to maybe three (a conversion of some nine percentage units) which means they get more gas out and earn more. As time goes on, these systems are getting more sophisticated, still.

Which type? Well, it is certainly a matter of a sound business plan and the payback period – the longer that is, the greater the financial risk. A critical factor in the business plan is the feedstock and dairy or pig slurry are not a very high gas yield. So, to get gas yield up, farmers, particularly in Germany, have been growing crops, notably maize, to add to the mix. Now, just consider that, according to UN figures, in a modern, comparatively efficient, USA factory, one tonne of nitrogen nutrient (close to three tonnes of ammonium nitrate) typically takes 21,000 kWh (yes, no mistake, twenty one thousand units) of electricity to manufacture and deliver.[108] So we take some of that, spread it to land after cultivating and planting seeds (burning diesel), harvest it and transport it (burning diesel), store it and put it into an AD plant, to get a fraction of the electrical energy out. That is environmental insanity.

So, if you need to bring down the pay-back period, it brings us back to earning cash from importing other materials. If they are controlled wastes, and if there is a gate fee, that can significantly add to profits and reduce pay-back period risk. It also brings more nutrients on the farm so reducing purchased fertiliser costs.

---

108. Gellings, CW & Parmenter, EC: *Energy Efficiency in Fertiliser Production and Use*, *Efficient Use and Conservation of Energy Volume 1*, UNESCO, 2004

Remembering the energy cost of manufacture of mineral fertilisers, we cannot afford the energy and pollution caused by manufacturing all the nutrients we need to feed the world. Ultimately, by one route or another, we have to put wastes back on the land. If AD is one of the routes to be used, then doing the homework, understanding the technology and applying it to the individual situation will help reduce the negatives. Oh, and we can still have the electricity or the gas for heating.

Thermophilic Aerobic Digestion – TAD
Basically, this process is blowing air through the mass of 'waste'. This can be done outside with materials such as pig manure in a lagoon in order to reduce odour. It can also be done inside which will cost very much more. The indoor route is useful for treatment of potentially high risk materials such as animal bi-products, or high odour materials; in both cases, total enclosure gives potentially total control. Blowing air will raise temperatures in suitable materials (those with enough nutrient concentration for feed the micro-organisms) to above 70°C which will give adequate pathogen control in most circumstances. With some feedstocks, up to 100°C can be reached (although the energy cost of blowing the air may become very significant). If extracted air is put through a carbon filter, then a very high degree of odour control can be achieved.

TAD will not produce methane and, therefore, is of no value in producing electricity. However, it has three very significant advantages over AD.[109] The equipment is very significantly lower cost. The output liquor is of significantly higher value in agriculture and horticulture because it has maintained nearly all the organic carbon, the nutrients in low-leachability form and has low odour. Thirdly, there is no explosion risk.

Hybrid Digestion

Some of the new technology, involving *aerobic* (to get the temperature up), *anaerobic* (to take 60% of the potential methane off) and then *aerobic* (to get the odour out) does appear to be promising. The idea is to try to combine the best of both worlds using TAD to start the digestion process and get the temperature to rise by either pumping air into the liquor in the process and getting a temperature rise entirely by natural process, or by using added heat to 'kick-start' the process; this may well be derived by burning some of the methane produced in the process. The second stage would be AD to get maybe 60% of the potentially available methane out, and the final stage would be aerobic and would reduce odour and leave some organic carbon.

The aim of these processes is two-fold: to reduce pathogens and reduce odour. Generally, the higher the temperature during the TAD phase (up to 60°C or so), the more likely it is that reasonable pathogen kill and odour control will be achieved. However, there is some evidence that best pathogen control will be achieved at between 55 and 60°C. Also, as temperature rises, so does oxygen demand rate, and lack of supply will result in a move towards anaerobic conditions with a consequential risk of offensive odour.

Take a look at the table which shows a comparison based on a subjective scoring principle. In a way, it is a bit trivial but it does show the value of the hybrid technology. TAD is very much cheaper in capital cost. TAD is much more flexible in terms of feedstock and ease of process management. The TAD final product is much more attractive agriculturally, because it is lower odour and high in aerobic micro-organisms with less crop disease risk; so, the market for the output is assured. *However*, TAD will not produce electricity directly in the same process and this is where AD scores. What the hybrid does is retain the advantages of TAD and gains the renewable energy output advantage of AD while discarding the disadvantages of AD. However, most of the organic carbon has gone (into methane) and is then not available for improving the soil and not to build the bio-active carbon sink (BACS).

Demonstrative scoring for TAD, AD and the new Hybrid technology

Characteristic	TAD	AD	TAD/AD Hybrid
Capital cost	100	20	75
Feedstock flexibility	100	40	100
Ease of process management	100	40	100

---

109. Butterworth, B: *Busting the Myths*, CIWM Journal, June 2015, pp28–30

Characteristic	TAD	AD	TAD/AD Hybrid
Renewable energy	0	100	100
Agricultural product	100	60	60
Total score	**400**	**260**	**435**

Perfect. Well, not quite. As the children in the car ask on a long journey, "are we there yet?" ...and the answer is, as usual, "nearly!"

The idea of the hybrid digestion, AAAD (*Aerobic* to get the temperature up, *Anaerobic* to get the gas out, *Aerobic* to get back to aerobic output and low odour, then *Digestion*) has been around for some time but has not swept the world. The potential advantage of the TAD-AD hybrid is that it is based on a tried and tested design of a TAD plant which has been dependable in process and outstandingly successful technically. This plant has a methane extraction process built into the last TAD tank. That is, quite simply, an ability to turn the air off, allow anaerobic activity to build up, take off the methane, and then turn the air back on to get back to aerobic output.

Summary on On-Farm Digestion and Bioprocessing
No doubt, there will be continued development of bio-digestion processes. Some of these will be large scale industrial processes, situated off the farm. However, there is a specific advantage of farm-based processes where the output of the process could be used as a fertiliser; farms have land and understand the logistics of storage and spreading. Fertiliser in terms of phosphate is running out and nitrogen takes very large quantities of energy to make it. So, logically, this returns to a fundamental theme of this book; recycling to land is vital to global food production and human survival. There are two basic types of bio-process which will continue to form the basic farm process; Aerobic processing by composting (in the open or in-vessel) and in the liquid process of TAD, and by anaerobic digestion or AD.

Output Values
At the time of writing, it appears that the UK government will continue to support 'renewable' energy production from AD. In that case, the current status of AD and how it can be used is worth a summary.

There is no doubt that, because of government subsidy on 'renewable' energy, AD is potentially an earner of significant profit. Beware, however, of those who peddle the AD myth of added value in the liquor made from farm-produced manures. The liquor is potentially *less* valuable than the manure from which it came for two very simple reasons. Firstly, the efficiency of the process, and the profits it might earn, depend on turning the organic matter in the manure feedstock into methane and that means there is less to form humus in the soil when the liquor is spread to land. Even people with the most basic knowledge of soil accept humus as part of fertility. Secondly, the liquor is often claimed to have nutrients which are 'more available' (what happened to the nutrients which are 'unavailable'?). Availability is measured in the lab by testing for solubility which means that those nutrients will leach out easily. Logically, unless something extra is added, AD liquor cannot be worth more than its manure feedstock (just hang onto "unless something extra is added"). Soluble materials can and do wash out in rain or irrigation. It is relevant to repeat an earlier figure, modified as below, which shows how soluble nutrients get into the plant or how losses take place. It is the same for ammonium nitrate and the way nutrients get into the plant in this 'available' system is basically in solution, i.e. it is hydroponic. The way to minimise losses of nutrients (it is not possible to *eliminate* losses) is to apply little and often during the growing season.

There is a way to almost eliminate losses and that is to put the liquor onto a compost heap. Forgive the repetition, but from this new angle; what happens here is shown in the figures in *Chapter 12* above, and sometimes called *eco-mimic* because it is the natural ecosystem way of doing things and is organic based. The micro-organisms in the compost heap will digest the soluble nutrients and turn them into their own bodies in enormous numbers. When they die, the breakdown products form a complex black tar we call *humus* which is insoluble in rainfall or irrigation. What the soil fungi (the *mycorrhizae*) do is feed at one end on the humus and the other end either crosses the root hair into the plant or wraps round it like the placenta of a mammal. The mycorrhizae actually feed the plant through a closed conduit. That is why natural ecosystems do not leak their nutrients quickly (the old 'organic boys' were right all along but they did not know why). Nevertheless, the nutrients are still available, on demand, by the crop roots. How do we know this? Well, think of the Fens, the most productive soils we have, made up of a very high fraction of organic matter, most of it as humus and yet the best crops in the country.

HOW SOLUBLE NUTREINTS ARE LOST

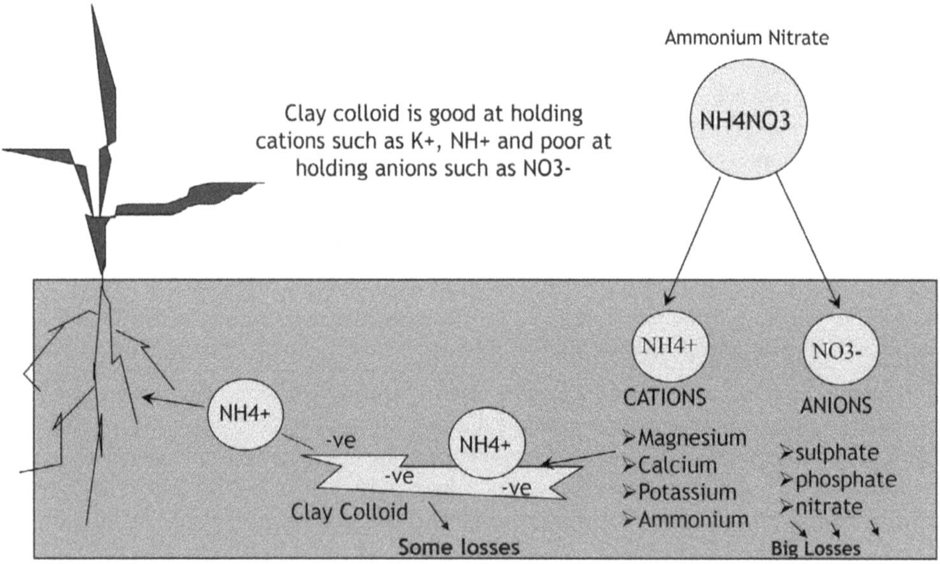

Ammonium Nitrate

NH4NO3

Clay colloid is good at holding cations such as K+, NH+ and poor at holding anions such as NO3-

NH4+   NO3-

CATIONS	ANIONS
➤Magnesium	➤sulphate
➤Calcium	➤phosphate
➤Potassium	➤nitrate
➤Ammonium	

NH4+   -ve   NH4+
-ve   -ve
Clay Colloid
Some losses   Big Losses

When mineral fertilisers such as Ammonium nitrate are applied, the cations are held in the soil colloid "bank" which also holds water. However, rain will take nearly half of the nitrate into groundwater.

Fig 14.6: How Soluble Nutrients Are Lost

Bio-Gas for Domestic Fuel and Local Small Transport

In Europe, mainly because of regulation, small digestion plants are generally not used. However, in other areas, particularly in Asia, and specifically in China, AD has been widely used at a domestic level for decades, mainly for cooking. Household wastes are used to produce gas from small, below ground, digesters. These are made using local skills and materials. Gas is now sometimes sold in plastic bags. This gas is also used to run small engines and transport.

Combined Heat and Power

All digestion processes produce heat energy, as do most biological processes, but it is often low grade heat, i.e. not much higher temperature than the ambient. However, some processes produce more and, in particular, when AD produces methane which is used as a fuel in an engine to drive a generator, then the engine will produce much heat at a usable temperature. In some cases, this can be used productively, such as in a dairy unit or glasshouse, and this may be financially justifiable in its own right. It may also eligible for a subsidy where governments make that available to encourage 'renewable' energy production.

Of all these processes. Methane is a clean burn – it produces carbon dioxide and very little else.

Algae Production and Biofuels from Algae

The oil companies in particular have become increasingly interested in algae, such as cyanobacteria and spirulina, simply because they are green plants and they may be manageable in liquid systems which can be mechanised. This explains in one short sentence why algae could be useful in the oil industry. In theory and in trials, it is possible to produce engine fuels from algae. They will also do what the green leaf in farming does; remove carbon dioxide from the air and pump oxygen back in. It is not the purpose of this book to examine the possible route for culture of algae to lead to liquid biofuels. The researchers, however, may draw from agricultural experience. We know from high-tech inputs into the genetic manipulation of crops such as maize, coupled to carefully designed cultivation, that we can get close to total utilisation of the available energy from the sun by restructuring the way leaves grow, and now in the manipulation of the photosynthesis process itself at molecular level, so that they can 'mop up' all the available energy from the sun. If we have a layer of algae, the same rules apply; there is a limit to how much sunlight falls on an area and algae don't grow and fix the sun's energy in the dark. There may be a very long way to go before these systems can be managed in commercial practice. Algae are certainly interesting and worth pursuing. For the here and now, however, we know that we can reclaim deserts and uplands with 'wastes' and produce both biofuels and food by relatively well-established normal agricultural techniques. We know they are safe. We know they work.

Chapter 15: Which Wastes

Background

John Lawes, generally regarded as the 'father' of the mineral fertiliser industry, did not start his first superphosphate factory until 1843. Progress was initially slow but 'artificial' fertilisers eventually became cheap to produce, very concentrated (simplifying transport and spreading), and they made dramatic, visual and rapid differences to crop appearance and yields. Once large scale production became possible, it took only fifteen to twenty years to sweep the world. These 'artificial', or mineral, fertilisers are a comparatively recent invention. Until they arrived, farmers used mainly locally available wastes, but also guano (bird and bat 'droppings' or *faeces*), transported across the oceans by the great sailing ships, for this same purpose of fertilising crops. When locally available wastes, such as cotton *shoddy,* leather trimmings, slaughterhouse wastes, were available, they were used as fertiliser. Grain was imported from overseas and fed to livestock, and the urine, dung and soiled bedding (muck) was used for the same purpose. So was local human sewage. Waste has always been recycled to land, ever since *Homo sapiens* began to wander and even before he settled. It is important to understand that everything originally comes from the land and it will eventually go back to this same land. Given sufficient dispersal rates and time, nature will have its way.

Specific Materials and Agricultural Values
Generally speaking, 'green waste' from domestic gardens is thought to make ideal compost, but this is not necessarily so. Garden waste from high density housing built maybe within the last five years and the waste collected in, say May and June, will be high in nitrogen and likely to be of high value in composting and to the land onto which it will be eventually spread. However, garden waste collected in, say, October and November, from detached houses built fifteen to twenty years ago, is likely to contain large volumes of cupressus and, if much of this is trunk timber, a lot of energy will be used to shred it and there will not be much nitrogen in the resultant compost. There is also likely to be a problem in the slow breakdown of the resin in the trunk timber from conifers. If this material is not composted with both another nitrogen source and significant moisture added, then it is quite likely that crops 'fertilised' with the resultant compost will be negatively affected.

Currently, British farmers annually use millions of pounds worth of mineral fertilisers, with the basic raw material almost entirely imported. All sorts of 'wastes' could, technically, replace most or all of that. Could it be done safely and economically? Would recycling to land be at a lower cost and more sustainable than current separation and collection routes? We may have become too obsessed with sophistication and centralised industrial processes when the most sustainable route has been in use for centuries. The old boy in the country cottage garden; he often lived to be over a hundred years old and his garden soil was black with organic materials. He never threw anything away (even what we politely call *night soil!*) He recycled everything he did not need, to the soil in his garden.

He got all the immune stimulation he needed, at all the right levels, all the minerals and all the vitamins. Mostly, it worked incredibly well.

Self-Contained Farming

On the face of it, and logically, a 'mixed' farm ought to be able to support itself without additional fertiliser input, provided imports of nutrients balance export of nutrients in exported crops and animals. If the crop residues and animal manures are all composted down during or at the end of the year, then it is all just going round again. All of it, the trace elements, the magnesium, the sulphur, the calcium (lime), the potash, the phosphates and even the nitrogen. This is even more likely to be true because, in the UK (more in tropical, high rainfall parts of the world) every hectare of land gets nearly two kg of nitrogen fertiliser out of the rain every year. This comes from lightning flashes in thunderstorms – most of which you will never see or hear. The searing temperatures of a lightning flash are high enough to force the nitrogen gas in the air (that nitrogen is normally inert and will not react with anything) to react with the oxygen in the air, forming nitric and nitrous oxides. These dissolve in the water in rain to nitric and nitrous acids. These are, of course, very dilute as acids but the nitrogen is in the form of a fertiliser which the plants in your garden can use.

However, farming usually removes significant quantities of these nutrients in exports. Further, and more insidiously, some of the nutrients, particularly nitrogen, will leach out in the rain. Not only that, the more the soil is cultivated, the more the organic matter will oxidise and go off into the atmosphere or groundwater. This is particularly true of the nitrogen in the soil. Nitrate fertilisers will leach out in the rain or during irrigation very easily, even on a clay soil, more so on a sand. Typically, nearly half the fertiliser applied, if it is mineral nitrogen, leaches out. If the nitrogen is in the form of the proteins in humus, it will not leach out, but if you cultivate the soil, it will tend to oxidise. As that happens, if the soil is moist, the nitrogen will turn to nitrates and nitrites and be easily lost into the groundwater. In very hot, dry weather, it may go off as ammonia into the atmosphere.

It is over-simplistic to conclude (but there is a point here) from this that the worst thing a farm can do from a nutrient maintenance point of view, is cultivate. The alternative is to mulch and let the worms do the work.

Values of Wastes as Feeding-Stuffs for Farm Animals

Generally speaking, if there is carbon in a big molecule, then compost heaps and ruminants may well be able to use that material. As a guide, if we could eat it, so could a farm animal. More than that, and importantly, ruminants can easily digest cellulose and can, therefore, eat large quantities of fruit and vegetable wastes for food processing factories. Both cattle and pigs can eat dairy product wastes. The question of how far this can be pushed with respect to wastes is not an easy one. For example, in theory, it is quite feasible to feed bioglycerol to ruminants but there are two questions to be answered. Firstly, bioglycerol comes from biodiesel manufacture and if the feedstock for that were used cooking oil, then the bioglycerol is a *controlled waste* in the UK and regulations of the EU (the European Union) could not be taken to a farm for feeding. If the feedstock were virgin crop oil, then the bioglycerol is not a controlled waste (i.e. defined as 'not waste' under those regulations and could, legally speaking, be fed. This raises the question of whether there is significant residue of methanol left over from the process (as the biodiesel manufacturing process is methyl esterification based on methanol with a catalyst). Small traces of methanol probably would be metabolised in the rumen but there is little known on how far that can be pushed. Alternatively, the bioglycerol can be composted quite easily.

Municipal, Industrial, Business & Domestic Wastes
Generally, municipal authorities are a significant collection route for 'wastes' and, as far as the European Union is concerned, Brussels has dictated progress by municipal authorities to collect and recycle. The main thrust of much of the regulation has been to direct separation at source, i.e. at the domestic producer household level or at local recycling sites. This 'source separation' of *municipal solid waste* (MSW, otherwise known as bin rubbish or garbage) involves very significant expense and produces poor quality separation (householders are not very consistent in separation) which limits remanufacture and delivers an unacceptably low level of safety for recycling to land. This problem of cost and less than satisfactory quality of output will result in developing alternatives which involve collection of whole, unseparated domestic waste. Source separation will fade out as these alternatives become available. That advance will, inevitably, be inhibited by regulation which always fetters any entrepreneurial progress.

Sewage has, historically, often been a municipal responsibility and, therefore, this source of waste has been identified, exploited and regulated. It still has a stigma attached to it in the public and politicians' eyes for, perhaps, understandable reasons and, therefore, there are from time to time, moves to make it 'disappear' by gasifying (*pyrolysis*) it or some other way of extracting energy by incinerating the carbon in its organic matter. This, of course, is an enormous waste of its potential soil value as a very effective and safe organic manure due to its ability to assist in reduced use of crop protection chemicals.

In Europe generally, most businesses and industries are significantly behind municipal authorities in actually following the restrictions and pressures imposed by the regulators. Nevertheless, because of rising landfill costs, all businesses will become increasingly active in managing their wastes better. It makes sense that raw material should be recovered to reuse in profitable production if possible. The key word is 'profitable' simply because handling, processing or doing anything involves resources and has financial implications. The source of money involved depends on the general business environment of prices of raw materials, regulation and related taxation. It is also relevant to think medium and long term in that the climate of what is 'acceptable waste' is changing.

Recycling industrial wastes to land is a real technical option limited mainly by regulation. But now is the time to reconsider recycling of a whole range of materials. Everyone in the waste business knows that it is possible to recycle green wastes (from gardens) to land. However, there is a limited supply of that and industry is facing dramatic rises in gate fees and restrictions in going to landfill or to high-tech processing. The plain truth is that the most high-tech processing operation yet designed by man is insignificant in its sophistication and safety compared to the complexity, thoroughness and safety of a compost heap and a fertile soil.

Zinc is one of those 'heavy metals' in garbage and also in sewage and biosolids where it comes mainly from what women put on their babies' bottoms (for protection against 'nappy rashes') and on their own faces (in cosmetics). The metal is also used in industrial processes, especially *galvanising* and effluent from such industrial processes may be significantly polluted with the metal. Zinc is also one of the trace elements essential in the healing process in the human body and is often added to animal feed in a mineral and vitamin mix. So an informed balance needs to be kept.

The absence of concentrated industrial areas, as is often the case in developing economies, may reduce the risk of heavy metal pollutants in garbage. Where light industry is integrated into domestic areas, this may change the risk: in this case, heavy metal monitoring is likely to be more important. In any case, we have the technology to process urban and many industrial wastes safely. We also have the technology to plan what should be put on the soil, monitor the soil and protect it long term. Give farmers the technological support and the job can be done. There is one further area of input which is likely to be wise and high-tech. It is the long term monitoring of land if it is to be used for food production. The living universe of a productive and biologically active soil is continually changing its nutrient balance and is quite capable of locking up or dumping elements into the groundwater. Too much sodium can, for example, be pushed out of the soil by adding a calculated quantity of sulphur in the right form. This technology was initially developed by Dr. William Albrecht[110] and later by Neal Kinsey.[111] It is now more widely used and it does give the safety management which is needed long term if wastes are to be the main source of nutrient supply.

The micro-organisms in a compost heap are mainly bacteria, actinomycetes and fungi. The soil is of a similar composition but usually with an emphasis on fungi. The numbers in a cubic metre of active compost runs into trillions and the diversity is highly variable (depending on type of material present and the stage of breakdown).

Feeding the Micro-Organisms.

The micro-organisms in compost and land are different from us, of course. However, in terms of dietary needs, they are remarkably similar. They are different in that they can tackle much bigger molecules than we can. Just as we might put sugar (a 6-Carbon molecule) on our cereals for breakfast, the micro-organisms can tackle the lignin in wood (with an almost unlimited number of carbon atoms in the same molecule) and they never stop eating – 24/7.

As an example of this capability, it is interesting to look at one of *Land Research*'s farms (in a consortium of farms which recycle wastes to their own land). The example farm took wood from local municipal and landscape gardening sources. Cellulose is a polymer made up of long chains of maybe 3000 monomers (groups of five carbon atoms in a ring). So there are around 15,000 carbon atoms in one of these chains. Lignin is made up of cellulose chains lying next to each other and joined by cross-linkages. This means that xylem (the 'wood' behind the bark) in a tree may be almost just one molecule – everything is joined together. The number of carbon atoms joined together is, obviously, enormous and almost so big that it is beyond ordinary number-comprehension. This farm also took in PVA – polyvinyl alcohol as a hot/warm liquid (it solidifies on cooling). As a liquid, it is comparatively easy to disperse it through a large mass of compost

110. Albrecht, WA: *The Albrecht Papers, Charles Walter Books*, 1919–1970

111. Kinsey, N: *Hands on Agronomy, Acres USA*, 1999

(preferably active and, therefore also hot). This is a big carbon chain alcohol with many thousands of carbon atoms in each chain. Even so, lignin, the basic molecule of wood, is even bigger with many cross-linked chains. As soil and composting micro-organisms can crack lignin, it follows that they can crack PVA. The BOD (biological oxygen demand – a measure of pollution potential) is enormous, so it is necessary to keep turning the compost (to keep the break-down activity active) but, as a result, the bio-activity rises rapidly and the process is faster. It helps make very good compost. PVA is a plastic; so we can and do compost plastic provided the format allows the micro-organisms to digest it.

So, compost heaps can break down almost any organic molecule – 'organic' meaning molecules based on carbon chains. Given a balance of other necessary food molecules, they will crack 100% of the carbon chains in the feedstock in time. Furthermore, it does not really matter if there is any residue before spreading to land, provided that the soil is biologically active (meaning that it already has a reasonable amount of organic matter in it). Nevertheless, solid and dense plastics are an exception in the practical sense simply because digestion will, in most cases, take too long.

To say that the carbon chains will be broken down is not the complete story. The micro-organisms make new carbon chains in the form of hydrocarbons, carbohydrates and proteins. When the micro-organisms die, these molecules form a dirty black tarry substance (sometimes referred to as DBS or 'dirty black stuff') – initially very similar to crude oil. This product is actually what most of us call *humus*; it is the material which gives soil its black colour and its fertility.

Generic values of materials and values to land and crops. Within the context of this text, then, there are a few generalities which give some guidance. If the waste has a lot of carbon in it, then it can be composted or burned (for energy recovery). However, if the carbon is present in large, digestible chains in the molecules, it is valuable as an energy source for the micro-organisms. If any useful agricultural crop nutrients are present in the waste, it usually can be composted and recycled to land. These nutrients are the major ones of nitrogen, phosphates and potassium salts, followed closely by calcium, sulphur and magnesium and a whole range of *trace* elements which include most of the heavy metals (that regulators generally shy away from).

Evolution has created an ecological balance which can not only tolerate most things, but actually needs most things. For example, as you read this, your blood will contain around 10ppm (parts per million) of arsenic. However, too little or too much arsenic would result in illness and probably death. The same is true of many other elements and that is why they are referred to as *trace* elements.

The suitability for recycling to land for agriculture, energy crops, forestry, or land reclamation depends on the composition and dispersal rate of the material in question. This area of technology of environmental management is known as *dispersion technology*.

There is one other possible way of judging safety; if digestible by humans, it is, possibly/probably, digestible by the micro-organisms in a compost heap. Humans have difficulty in digesting the cellulose in garden wastes but ruminants do not. There is a remarkable similarity between the micro-organism universe of a compost heap and that of the rumen of cattle, sheep, goats, antelope or any other ruminant.

Logic says that the high cost of source separation, so favoured by the EU and the UK governments, will eventually be superseded by more dependable, and less expensive, recycling routes.[112] For example, members of the public may be educated to put a small NiCad battery in a separate bin, but not all (100%) of the people will do that. What happens in practice is that the resultant compost is spread out far enough to reduce the risk of pollution from that battery to an acceptable level. That can also be achieved without the cost of source separation and the technology for central separation is there, ready and waiting.

Moisture in the Composting Process

All composting operations are driven by micro-organisms. And nearly all life as we know it is dependent of moisture. Without moisture in a compost heap, it will result in preservation rather than processing. Composting garden waste without enough moisture will produce a 'woody hay'. It may be dark coloured and friable but the process of pathogen kill, of immobilizing the soluble nutrients, of killing weed seeds, will all be less than complete. Sadly, some of the commercially available, bagged 'composts' available in garden centres and described as 'peat free' may be of this sort of character.

Having recognised that, there is a practical difficulty in deciding what level of moisture to aim for to achieve the best process. Firstly, materials delivered to a composting site may have too little or too much but they have what they have; adding moisture may be easy enough in principle if it is available (in hot climates, it may not be), and removing moisture could be difficult (such as being so wet that the compost activity has difficulty in starting and generating the temperature that will drive off the excess moisture). It is also the case that under EU law, and particularly in the UK with the Environment Agency, the addition of *potable* (i.e. drinkable) water from the main might not be questioned but the addition of liquid wastes would, in many or even all cases, be prohibited. There are many industrial wastes which could and should be recycled by addition to composting processes and the nutrients in them, however dilute, be recycled to land.

There are two important functions which will dictate where the optimum is; aeration and adequate moisture for process. Firstly, if the input material for composting is well structured, such as with coarsely shredded and woody garden waste, then there will be enough air for the process to start and to build up heat. Furthermore, the windrow or heap will be able to breathe, much as sub-marine snorkels, letting heat out and new air in.[113] From a farming point of view, a range of particles up to 100mm in diameter will give a significant amount of flexibility as to how often the heap needs turning. If there are large particles, then the heap will, in a wide range of circumstances, breathe on its own, without human intervention. As particle size decreases, then that flexibility is lost but the process will progress at a faster rate. Also, where there is a significant proportion of large particles, the material could be very wet and the process still progress satisfactorily. If, however, the material is mainly something like lawn mowings or white onion flesh, then it will slump very easily and run out of air very quickly and go anaerobic and smell offensively.

112. There are many references in many places including on the web, for example at Washington State University, Cornell University and others

113. Butterworth, B: *How to Make On-Farm Composting Work*, *MX Publishing*, London, 2009

So, where possible, a mixture of materials will give optimum process control. Commonly, this will have moisture contents in the 40–60% range. If the material has good air spaces connected to external air, that moisture content could be as high as 70%, but, especially when poorly aerated, above this temperature, there will be an increase in odour risk. As the moisture drops below 40% there is likely to be process slow down.

There is one area where moisture is in significantly greater demand and that is with some industrially-prepared materials such as MDF (medium density fibreboard).[114] Fibre boards can absorb several times their own weight of water before expansion takes place which disintegrates the board and allows the composting process to proceed.[115]

This question of moisture in the compost process will be referred to again below with some test figures.

Wastes Which Are Suitable and Probably Safe

Repairing the loss by harvest, erosion or leaching is relatively uncomplicated in apparent principle; put nutrients and organic matter back into the soil by importing waste, preferably organic matter, from outside the farm.

Green Wastes: including lawn mowings, bedding plant residues, hedge trimmings, fresh and cooked vegetable waste; all good sources of organic matter that will help to make humus. Green leaves especially valuable because they contain nitrogen. Lawn mowings in the spring are the 'dynamite' of composting with much nitrogen and moisture. On their own, they will use up the oxygen quickly and 'slump' down into a wet, soggy mass and are likely to go anaerobic and smell. Mix mowings with other material and move it more often. In the autumn, remember that woody materials have much less nitrogen in them.

Kitchen Wastes: *Stop!* There is a legal problem in the EU and many other countries. If it contains meat residues, or might do so, then that material is covered by the *Animal By-Products Regulations* (ABPR) and that means composting or other treatment in isolation – in a building of some sort, sealed off from birds and vermin. However, if it can be handled, inside or out, then the micro-organisms in the soil and the compost heap are very much like ourselves in terms of nutrient needs, but they can, given time, digest almost anything. Below is a list of common kitchen wastes and their values.

Uneaten food after a meal: Vegetable waste as above, meat good (contains much nitrogen) but chop it up as it is slow to disintegrate.

Paper and cardboard: Not the best but by no means useless. Paper products are mainly carbon which means that the micro-organisms can use it for energy and as the basic building block to make humus. However, to do that, it will be necessary to also feed some nitrogen, in particular, and the full range of other plant foods including trace elements. These can come from the materials above in this list but don't over-do the paper. How much is 'over-do'? Well, it is partly experience but the technology will give you the framework (see *Chapter 14*).

---

114. Butterworth, B: *With or Without MDF*, **Far Eastern Agriculture**, September/October 2003

115. Butterworth, B: *How to Make On-Farm Composting Work*, **MX Publishing**, London, 2009

Ash from fires: Great if it is wood ash; that will give much potash and a good range of minerals (or trace elements). If it is coal ash, that will yield up the minerals but very little potash. All black coloured ashes have another effect (see next paragraph).

High Carbon Ashes, Charcoal: If the ash is very dark, really black, then it may have another value with a long term effect.

Studies of black soils (called *terra preta*) in South America[116,117,118] have put another angle on carbon and its value as amorphous or activated carbon, usually (it is thought) derived from charcoal. This charcoal is sometimes now trendily called *biochar*. Terra preta soils show a very high degree of sustainable fertility which, in areas of high temperature and violent fluctuations in soil moisture, might at first appear surprising. Initial studies on *Land Network* farms indicate that the carbon in printing inks appears to be of similar value. The carbon is capable of holding onto nutrients at a level similar to humus. However, humus is easily and rapidly oxidized under conditions of high temperature and moisture but the pure carbon is not. It may be that there is potential for the wider use of carbon in soils in ways that we do not yet fully understand. An article by Michael Marshall[119] in *New Scientist* looked at *terra preta* soils and, from traceable contents of fish bones and seeds, concluded that the resultant black soils may, at least in part, have been the conscious result of waste management by a fairly intelligent population.

So, dark ashes may be beneficial, especially in terms of holding nutrients long term.

Animal manures: Great stuff with plenty of food for the micro-organisms and a fresh injection of its own micro-organisms (which may speed up the whole process). Horse manure is a useful source of potash. Poultry droppings are very rich in nitrogen. Cattle manure is good in phosphate and nitrogen. *Land Research* farmers have even handled elephant dung. Yes, and the dog and cat can contribute, too. With all of these, because they potentially carry human pathogens, getting the temperature up is important, so do use a thermometer (note that, in regulatory terms, faeces from animals that might have eaten meat are, in the UK, covered by the ABPR and cannot be composted out of doors).

Crop Nutrients in Animal Manures

Manure	Likely %			
	Moisture	Dry Matter		
		N	P	K
Cattle	75–85	2	1	0.5
Pig	75–85	4	2	1
Chicken	60	6	6	3
Sheep	60–70	4	2	1

116. Berg, B et al: *Plant Litter, Decomposition, Humus Formation, Carbon Sequestration,* **Springer**, 2003. This is one example of many hundreds of references to organic matter degradation and formation of humus in the academic literature and on the web

117. Mann, CC: *Our Good Earth,* **National Geographic**, September 2008, pp80–107

118. Butterworth, B: *The Straw Manual, Spon*, 1986

119. Marshall, M: *Finding the Real El Dorado,* **New Scientist**, Jan 19 2019, pp26–29

Values of manures vary from species to species, with what they are fed on and whether there is bedding with the dung and urine. Some guides are as follows and these are without straw or other bedding;

Animal bedding: Cat litter may be made from volcanic ash which is inert and will add to the structure of clays. Sawdust is mainly carbon (useful for humus production provided there is at least some droppings and urine). On this latter material, in old Victorian gardens, the old gardeners would encourage the apprentice gardeners to urinate on the compost heap. Human urine (and of other animals, too) contains uric acid, which contains nitrogen. Urine is (because of the acid) mildly anti-septic; so it is unlikely to carry disease if it is fresh urine (best not to leave it hanging about – get it into the compost heap).

Carpet shredded to a 'fluff': Great stuff, especially if it is from wool carpets. Even if it is from synthetic fibres, these will be very valuable as fibre s stabilise the physical structure of soils and allow them to breathe and manage water better (see later in this chapter below for a discussion on fibres).

Cupressus prunings: It is possible to compost these but they will process faster if they are shredded. All the coniferous trees, however, contain resins and these are very difficult for the micro-organisms to break down. Turn more often.

Wood: Wood is mainly carbon and will need chopping or shredding into fairly small pieces.

MDF: MDF is very interesting. It is, of course, predominantly wood. There is, generally, a lot of misunderstanding surrounding MDF, chipboard and similar process wood sheet materials. Such materials are present in significant quantities in many of our homes, offices and the confined spaces in which we live. The most common binder or glue is ureaformaldehyde. Urea is used in large quantities, world-wide, as a nitrogen agricultural fertiliser. The second part of the molecule, formaldehyde, *if it were separate and on its own,* is known to be toxic and potentially carcinogenic. The combined molecule is not toxic. Sodium is a silvery metal which dissolves in water and bursts into flame as it does so; if it were eaten, it would burn a hole though the body until it came out the other side. Chlorine gas is quite poisonous and a few whiffs will kill. However, sodium chloride (common salt) is eaten as a food.

Ureaformaldehyde: Ureaformaldehyde is used both as an agricultural fertiliser and has been used for many years as a slow release nitrogen source in many general potting composts, commonly sold in garden centres[120,121,122] The key phrase is 'slow release' which has significance both in commercially in growing and environmentally in reducing pollution.

So, if people tell you this material is dangerous, ask for their evidence and suggest that, if they were to be right, we all have much more risk from the furniture in our kitchens and offices. WHO (the World Health Organisation of the United Nations) has had a look at this and there is a clean bill of health.) Everything has risks but, as far as we know at

---

120. FAO lists many references on the use of ureaformaldehyde as a fertiliser

121. Bunt, AC: *Modern Potting Composts, George Allen & Unwin*, London 1976

122. Handreck, KA & Black, U: *Growing Media: 3rd Edition, UNSW*, Australia, 2002

present, any risk from MDF is very, very low. The risk as a fertiliser is much lower that the risk in a typical kitchen or office where it is likely that all of the furniture is made of MDF.

So, how can you use MDF in composting? Well, it is necessary to let the bugs get at it. One way is to shred it (which takes quite a lot of energy) or break it up a bit (say 150 mm bits) so as to leave a broken edge. For tree planting, relatively large pieces of MDF, say 150 mm square, can be buried, preferably after soaking in water until it expands, under permanent planting for, forest trees or land reclamation in preparation for a permanent planting. That will give a slow release of nitrogen. Alternatively, shredded material, or just broken up into 150 mm squares, can be mixed with some farmyard manure or active compost made from this or other materials (these will add some micro-organisms and 'seed' the process), then make sure it is quite damp and kept that way. A fifteen mm thick sheet will keep on absorbing water until it is fifty to 100 mm thick. Then it will break up and can be mixed in with the compost or spread direct to the ground. It is a really good source of nitrogen – so, until you are used to it, not too much in one place and allow moisture and time. Broken MDF boards, mixed 50/50 (approximately) with cattle slurry will break down reasonably quickly – but still maybe two or three times longer than if it were shredded (if time is not a problem, this route is, of course, less expensive on energy).

Plasterboard: Plasterboard is a sandwich of gypsum, which is calcium sulphate, between two sheets of paper. The paper is organic carbon. Plants need both calcium and sulphur. The calcium is especially useful in helping flocculate clays to give a better crumb structure, make them easier to work and plants will establish faster and grow better. However, the calcium sulphate is fairly alkaline and so use sparingly if adding to a compost heap. Avoid plasterboard with an aluminium foil backing; that will not break down. The same remarks apply to waste plasters – most of which will be calcium sulphate based. Better technically (although UK regulations, wrongly, prohibit this) to put all of these direct to the soil where broken up plasterboard will break down as it gets wet but it will look untidy in the meantime.

Old mortar: Old mortar from an old wall will have been made from lime (probably *quick lime,* which is calcium oxide) and sand. Crops need the calcium; it will help the compost, the soil and the plants. However, the mortar does need breaking up, preferably sieving, and remember how much it is diluted along the way onto the soil will affect the following plant growth. Calcicoles like it; calcifuges don't! Mortars are also alkaline and, therefore, go sparingly if adding to the compost heap.

Coffee grounds: Yes, the compost process and the soil will deal with this, too. Spread it out a bit or it will keep the micro-organisms up all night.

Teabags: Have you ever thought how tough a teabag is? You put it in boiling water and push it round violently with a spoon, even squeeze it with tongs, and even the couple of millimetres round the edge does not give in. How do they 'glue' those edges to stand that? Why does the paper not disintegrate and let the tea leaves out? The answer is that this is high–tech paper. There really is quite a lot of advanced technology required to design and manufacture a teabag. The base material is paper but the strength comes from the addition of a polymer plastic. Most gardeners will already know that teabags are slow to break down; this is why. Is it bad or toxic? Not a bit of it. The plastic is composed of just carbon and hydrogen. It will take time to break down but it will break down and do no harm. The fact that it is slow to disintegrate is a benefit because it adds fibres to the soil.

Carpets and Fibres in the Soil

New & Used Carpet
Regulators have been very slow, wrongly so, to allow farming to recycle carpet to land. However, gardeners have used carpet over many years to wrap round a compost heap to allow the process to breathe while letting rain in and reducing heat loss. These are all very important assets of carpet which is used as a carpet; just as a piece of material, still recognisable as carpet and used in pieces of a metre or two across. Gardeners have used the 'fluff' from a vacuum cleaner for years by adding it to the compost heap and spreading it amongst the other materials.

All of that is a really useful application for carpet and it will eventually break down. But suppose we accelerated that breakdown by tearing it up. What is its potential value?

The majority of carpets sold in the twenty years up to about 2015 in the UK have been 80:20 wool:nylon. The backing has moved from jute, or hessian, to polypropylene. From around 2015, polypropylene was increasingly used for the pile as well. The immediate response is that it is the wool which is valuable. Wool is certainly of value because it is protein and that contains nitrogen. However, there are other things of value. Look, for a moment, at the following analysis.

Now, here is an interesting thing; the biggest single component is chalk! All plants use calcium, the main value of chalk in soil. By the way, the adhesive will break down releasing energy to the micro-organisms. The nylon and propylene have real value and will be discussed in some detail below. So, carpet has a value in providing nitrogen, calcium and fibres.

Woollen Fabrics
Wool is pure keratin – a range of proteins. Proteins contain nitrogen in big molecules. Big molecules do not leach out, so this is slow release nitrogen. So, old woollen sweaters? 'Plant' one under a shrub or perennial. Alternatively put it in a compost heap; it will be easier to turn the compost if you cut the woollen fabric up a bit before adding it to the heap. It will be slow to break down but an active, warm heap will do it. It really does not matter if it does not break down completely.

The Value of Fibres in Soils
The last few paragraphs above have revolved around the value of fibres in soils. So, what is that value? It is important to remember that the basic physical structure of soil is composed of mineral matter which is generally regarded as not biodegradable. Most soils, as any textbook, and the Environment Agency's publication *Understanding Rural Land Use*[123] will also point out that the structure of good, stable, really fertile soils also depend on plant roots and 'organic' matter to create a matrix which gives further characteristics to the mineral matter in soils in terms of stability, gas exchange, moisture management, cohesion and erosion stability.

---

123.   *Understanding Rural Land Use, Environment Agency*, NE-1/100-3.SK-C-BEKC

**TYPICAL BREAKDOWN BY % WEIGHT OF
AVERAGE 80:20 WOOL-RICH CARPETS**

80:20 with pp backing		% of total	
facefibre	35%		
wool		80%	28%
nylon		20%	7%
adhesive/filler	45%		
SBR		20%	9%
chalk		80%	36%
backings	20%		
PP		100%	20%

80:20 with jute backing		% of total	
facefibre	35%		
wool		80%	28%
nylon		20%	7%
adhesive/filler	45%		
SBR		20%	9%
chalk		80%	36%
primary backing	5%		
PP		100%	5%
secondary backing	15%		
jute		100%	15%

Fig 15.1: Components in Carpets

SBR is synthetic latex, and PP (polypropylene)
fibres are not UV stabilised.

It can also be observed that many hydroponic crop production systems use vermiculite, polyurethane and many other synthetics as a growing medium to provide a matrix for root growth and a structure with a large surface area which the mycorrhiza can latch onto. These materials will normally be described as 'not biodegradable'. This is not strictly true in that all of these fibres be they 'natural' (such as lignin from plant roots) or 'synthetic' (such as polypropylene from carpet backing) are all biodegradable but the process will, fortunately, take much time, maybe years.

Now, as anyone who has any knowledge of growing plants knows, peat is used as a growing medium. *Organic matter* is regarded as providing stability and a range of growing qualities to soil. These materials contain cellulose, hemicellulose, holocellulose and lignin. All of these are big molecules, larger than the synthetics in the carpet mix. It is the fibres in peat which are the secret of what it imparts to soils it is mixed with or to a growing or potting compost.

Case Study
Some years ago, while working on the *Enterprise Initiative* of the DTI, I was working on the filtration systems of a fresh water fish farm in Suffolk. To operate, the farm had to filter water from the tanks where fish grew and were fed to do so – leaving un-eaten food and faecal matter, of course. The successful filters were large tanks filled to the brim with hollow spheres about the size of tennis balls with a skeleton of surfaces much like the segment skins of the orange. The spaces between the surfaces were critical to allow water to pass through even after the build-up of a slimy, brown coating of the surfaces. It took several weeks for the active coating to develop. That coating was active bacteria and fungi, much as the mycorrhizae in soils. It worked. I have little doubt that fibres in soils work the same way, except that the mycorrhizae in soil can digest humus (organic carbon molecules containing, N, P, K, and a full range of trace elements) and, through the mycorrhizal conduit (see *Chapter 12*), cross the root hair wall and feed the crop plants. While wool can be digested to produce a slow release feed of nitrogen to the crop, without leakage to groundwater, it does not matter from a soil structure point of view if the fibres are a long life plastic, such as polypropylene.

Composition of Wool and Other Carpet Fibres
Wool composition is of *keratin* proteins which, of course, contain nitrogen and, in the case of keratins, useful amounts of sulphur. In proteins of the keratin group, several of the amino acids contain sulphur and, of course, so do the amino acids which form some microbial proteins, plant proteins and animal/human proteins. These nutrients are clearly large molecules and therefore the nutrients are slow release and on crop demand.

Most people don't know this but, in most UK carpet composition, the most common material in the carpets is, as shown above, chalk which is used as a filler and stabiliser for the latex which glues the tuft to the hessian or polypropylene backing. Again, the chalk contains calcium which is a secondary, but very important, nutrient; second only to the major nutrients of NPK.

Jute used to be (and still is in what we will get for several years until currently new carpets get replaced) the most common backing and this is a fibre of plant origin, supplying carbon and energy to soil micro-organisms. Cellulose is, of course, mainly carbon molecules in long chains of 5-Carbon rings. Woody stems, as in jute, will contain larger molecules of hemicelluloses with these long chains joined by cross linkages. There may also be some lignin. Lignin contains many hundreds of 5-Carbon ring, cross-linked chains. Lignin is closely associated in woody materials with cellulose and hemicellulose. The point here is that these 'natural' fibres contain many hundreds of joined-up carbon atoms. Most people accept that these fibres will take some time to break down in natural environments of soils or compost heaps, but these same people rarely question how long degradation takes.

The *synthetics* are large carbon molecules – actually with smaller chains than in many natural fibres (see above). In the case of carpets, they are not 'stabilised' against degradation by ultraviolet light; so they are a little easier for the micro-organisms to digest. There is no question about breakdown and degradation; these do occur. The questions are how long this takes and, with this material, in this application, at these dilution rates in the garden, is a difficult thing to be precise about. Indeed, straw ploughed in may last several years if the soil is not 'breathing' to that depth. Whether slow breakdown is a good thing depends on circumstances but there are few or no known, likely or measurable deleterious effects. On the contrary, generally speaking, fibres are a positive part of soil structure.

In an 80:20 wool:nylon carpet, the tufts contain, usually, 10% of fine nylon filament fibres. Nylon does have a very small amount of nitrogen in its molecule. We do not yet have detailed knowledge based on long term experience but the technology indicates that these fine filaments, when composted, and dispersed, will undergo significant degradation in maybe weeks rather than months in the soil (years are very unlikely).

Polypropylene (Unstabilised)
The table below on the composition of carpets shows two analyses. The first is of jute-backed carpet – the majority of post-consumer carpet is like this. New carpets are like the second – for carpets which are 'post consumer' (taken out of the house when replaced by new carpet) we expect a progression towards this material over some years. Ten or fifteen years ago, there was little polypropylene used in new carpet backing – it was all jute (or hessian). Now, the situation has reversed. As these new carpets age, it might be expected that the percentage of polypropylene in 'waste' carpet will go up. However, there is technology which is developing which can separate the synthetic from the natural fibres and there is an incentive in that recovered synthetics can be worth over £400 (US$600) per tonne at 2009 prices. So, it is reasonable to think that the percentage of polypropylene may go down, rather than up, as the new carpets come up for replacement. What there is more caution about is how long polypropylene, if and when it is present, will take to break down under these circumstances. The soil science says that there is probably just as much advantage in mot breaking down, as breaking down.

Why all this on carpets and fibres? Well, fibres are an important part of the structure of soils. Cultivation will accelerate oxidation of these large molecules and the soil will breathe less, water will flow through less, there will be less water held in the soil in a 'healthy' way and there will consequently be more drought stress. Fibres make soils more productive and plants less subject to diseases. Erosion of soils by air and water is a significant factor in the loss of soils in all arable farming systems, all over the world and the UK is no exception. The UK Environment Agency recommends[124] the build-up of organic matter in topsoils so as to stabilise them against erosion, reduce surface run-off and thereby reduce the risk of flash flooding.

Municipal & Industrial Wastes

Just, for a moment, look farther than your boundary fence and look at the national picture. Currently, Defra (in its great wisdom) is encouraging local authorities to 'recycle' wastes through large industrial processing plants, i.e. *Energy from Waste*.[125,126] Whether you think that this is incineration by another name (which it is), and whether you think the energy gained means we save a small amount of burning fossilised fuels, is a discussion which appears to have been lost somewhere. What the route does do (because the installations the UK is going for are very large ones) is, without any doubt, centralised waste collection. This maximises trucks on the road and destroys organic matter which gardeners, nurserymen and farmers could grow flowers, food and biofuels with, so saving imports of mineral fertilisers and getting trucks off the road by proximity principle recycling. There appears to be a misguided obsession with large scale operations because of a blind assumption that there is economy of scale. It is interesting to note that comparatively small towns in Scandinavia may have two or more energy from waste plants. They are flexible and can change relatively easily with market conditions and they get trucks off the road. Heat is comparatively easily piped into local homes. Perhaps most importantly, the local people see them as *their* property.

Farming is closer to recycling on a *small is beautiful* or *proximity principle*. There are a lot of farms and if farmers do it, and the farmers, and the foresters, and the national parks, and all land users who currently use mineral fertilisers, yes, the pollution of groundwater can be eliminated and infringement of the standards laid down in the EU *Nitrate Directive* avoided.

Turning wastes, via composting or direct application, puts the organic carbon into the soil bank and the green leaf takes carbon dioxide out of the atmosphere and further adds to the carbon bank. Soil is a carbon sink.[127,128]

Wastes in the Next Millennium

The following piece is extracted from a paper published in *Resource*, the journal of the American Society of Agricultural, Biological and Environmental Engineers in July 1998,[129] and is reproduced here by kind permission of the publishers. It does, in fact

---

124. Butterworth, B: *Busting the Myths, CIWM Journal*, June 2015, pp28–30

125. Freyberg, T: *Gasification to Turn 200,000 tpa of Municipal Waste into Jet Fuel in Nevada, Waste Management World*, May 12th 2015

126. Butterworth, B: *Waste in the Next Millennium, Resource, Journal of the American Society of Agricultural and Biological Engineers*, July 1998

127. *Managing Municipal Solid Waste, EEA* Report No 2/2013

128. Butterworth, B: *Reversing Global Warming, Refocus*, September/October 2006

129. Butterworth, B: *Waste in the Next Millennium, Resource, Journal of the American Society of Agricultural and Biological Engineers*, July 1998

duplicate some of the information given elsewhere in this text. However, it did two important things at the time which remain important now. Those two things were firstly to crystallise a vision of farmers as a major global force in recycling urban wastes and, secondly, to show that recycling organic materials to land could make a significant contribution to the reduction of crop disease (the old 'crank' point of view of organic farming was, in the respect of crop disease at least, right).

De-centralised waste management to land offers a major new environmental role for agricultural engineers world-wide. New developments in Europe in spreading direct to land, composting andaerobic thermophilic digestion are opening new economic avenues for the recycling of waste. Advantages include recycling as being lower cost than disposal, a significant new source of income for farmers, reduced fertiliser costs, major reductions in rainfall/irrigation requirements, reduced crop protection chemical use and reduced nitrogen run-off.

What the Rio, Kyoto and Bali environment summits highlighted was the balance and, indeed, conflict between energy consumption and environmental stability. Consumer societies produce huge quantities of garbage and other wastes. The only factories big enough and sophisticated enough to cope are the sea and the land. The sea we cannot yet control adequately but managing the land is man's oldest technology. In the next ten years, the cost of logistics will push waste handling into an inflation leader. De-centralised waste management, turning waste into fertiliser, doing it on the farm, will produce a major contribution to environment management and changing farm incomes. There is indeed here a major new role for agricultural engineers.

High BOD Materials such as Bioglycerol or Fats

Fatty and oily materials are often available for composting as liquids. Some will be solids at ambient temperatures. The good news is that such materials are almost always based on large carbon chain molecules thereby presenting a lot of available energy, with the result that they can really activate composting process. These materials often command high gate fees as such a high level of activity is necessarily difficult to manage. Generally, the first problem is, as the designation of 'high BOD' implies, high *biological oxygen demand*. In practical terms this means that the heap will need frequent turning in order to avoid anaerobic and odorous conditions. Managing this needs a competent operator with some 'feel' for the composting process. Temperature is a useful guide and it may help in large operations or if skills are still developing, to use an oxygen metre. As a guide, if the temperature stays up to normal levels, not less than 50°C, and does not dip at all, the process is still aerobic. Regular turning in line with temperature, will maintain this aerobic condition.

Initiating the active process may be difficult and there are several factors influencing that start-up and the continuance of process. The nature of the absorbing solid material is critical. It needs to be only fairly coarsely shredded, in order to hold air. Somewhat in conflict with that is the nature of the viscosity of the added liquid. Some liquids may only be pumpable at well above ambient temperatures. They become more viscous on cooling, sometimes to the point of solidifying. Examples of this sort of material include PVA (polyvinyl alcohol) and, at low ambient temperatures, bioglycerol (from biodiesel production from virgin crop oil). To hold these materials in the compost heap is easy enough at the start as the material cools. However, if only a small amount of material is added to a large mass of active, very hot compost, then the liquid stays liquid and may simply run through material if it is too coarse. This situation can be very much worse if a large amount of liquid, especially bioglycerol, is added to less active, cooler, compost-in-progress. When the process re-activates, sometimes in two or three days' time, the temperature inside the heap warms up the material again and it becomes less viscous and may run out of the bottom of the heap again. Getting it right is largely a matter of experience in adjusting the initial quantity and its dispersal in the available solids, as well as in monitoring and reacting to developing conditions. The bottom line is frequency of turning and a simple water/leachate collection system. Generally, if material of the type which does solidify at ambient temperatures does leak out of the bottom of the heap, it will solidify quickly and not go into the liquid collection system; it can be swept/scraped up by mixing with compost. If it does not solidify, it will get carried into the collection system – another reason for designing a simple, open-to-the air, easy-to-pump or scoop-out system.

Liquids

Concrete is expensive in both energy and financial terms but the regulators often demand it under 'wastes' because of perceived risks. It is certainly true that large amounts of liquids added to not enough solids will result in the unabsorbed liquid running off and either being caught on the concrete or causing a pollution problem – hence the regulation of concreting under compost heaps in the EU. However, with just a little intelligence, matching added liquid volumes to the capability of the heap to absorb that quantity is not very difficult.

Generally speaking, making composts with solid materials but without the addition of some liquids does not work very well. Of course, if the materials are fairly wet at the start, it can be managed. However, in such circumstances, anaerobic conditions and the resultant odour is a greater risk and there will certainly be a need for careful monitoring and probably more turning.

Addition of liquids may be by spraying, preferably piping the liquid onto the top of the heap with a dribble bar (to avoid spraying liquid up in the air), by 'lagooning' on top of the heap or by what is best described as 'dunking'. The advantage of dribbling onto the top of the heap is one of control but it may, under some circumstances, mean that a person has to climb on top of the heap – and that is best avoided; the pipe is best positioned and secured using a telescopic loader. The lagooning method involves a compost heap of at least 3m depth and the use of a crowded loader bucket to create a dry lagoon on the top of the heap. The lagoon needs be maybe half to three quarters of the volume of the delivery tanker (depending on the viscosity of the liquid and the absorptive capacity of the compost in the heap). The tanker next pumps its load into the lagoon and this, with a twenty-five tonne typical load and truck pump, will take twenty to thirty minutes. The process is monitored for leakage but usually allowed some absorptive time. If there is leakage or after twenty to thirty minutes, the loader bucket is used to mix the two materials rather as in mixing concrete by hand with a shovel. It works quickly, safely and at low cost.

An alternative, the *dunking* method, is low cost and safe, ideal where a site has a run-off lagoon or trough in which a loader can work. In *Land Network* sites, such a trough, found at the end or at placed about one third of the length of the compost area slab is common. If liquids are pumped or dumped into that trough, then new and shredded solid materials for composting can, at delivery, be dropped into the trough to mop up the liquid. The wet solids can then be added to the compost in the normal way, allowing run-off to go back to the trough. Alternatively, if the trough is designed to be just a little wider than a loader bucket, the liquid can be scooped up and bucketed onto the compost heap. A further alternative is to pump the material and spray it onto the heap.

Where a tanker can be equipped with a large diameter hose which can be dragged over a large heap with a loader or 360° excavator–type machine, then very large amounts of liquids can be added and this was one of the reasons Land Research chose the *deep clamp* method of composting. With around three metres of absorbent material below the hose, this is the best possible structure/least risk which will retain liquid in the compost mass, and the easiest and lowest cost for mixing.

It is worth stressing three things with regard to adding liquids to compost. Firstly, the process is a biological process which depends on moisture. Secondly, the right amount of moisture is critical to the micro-organisms consuming the solubles and 'locking them up' into humus and making subsequent spreading safe. Thirdly, if the moisture is added as a 'waste', then it is recycling and financially rewarding.

Hospital Wastes

Of all the wastes produced directly by humans, faecal matter, urine and body parts are potentially of the most value for growing food. The nutrients and organic carbon are basic components of fertility. Yet we add copious quantities of water and dilute the sewage, and incinerate the body parts (producing carbon dioxide).

Historically, these wastes were seen as dangerous and to be disposed of carefully. The potential value of hospital wastes gives the regulators a real problem because there are risks of which the most obvious is that hospitals, by their nature, are likely to be potential sources of pathogens.

There is here, a potential classic example of finding a problem and a mirror image problem, then putting the two together in such a way as to produce not only a cancelling out of the problems but delivering a synergy to give an extra plus.

Despite the potential for pollution, hospitals produce significant amounts of wastes of many types, including many biodegradable materials containing significant amounts of organic carbon. There is usually a very significant cost of disposal running into many £millions nationally. Hospitals also spend significantly on energy, particularly, in temperate climates, heat.

In 2018 in the UK, there was publicity about the shortage of incineration facilities for obnoxious waste and body parts from hospitals. This, in turn, raises the question of the energy cost of incineration and the production of GHGs.

UK farmers currently (2019) buy around 1.6 million tonnes of NPK fertilisers pa, of which approaching one million tonnes is N and, according to UN-sponsored research, one tonne of N nutrient, made in a modern USA factory takes 21,000 kWh (yes, twenty one thousand 'units' of electricity) to manufacture and deliver, (mostly from burning fossilised fuel oil with a related CO_2 production). The farm-gate values of these fertilisers over £3 billion and it is all imported.

Putting the Problems Together

While some hospital wastes (hard plastics, reject scalpel blades, syringes etc) may not be technically suitable for anaerobic digestion, most do contain biodegradable carbon and would potentially be good feedstocks. However, there are certainly technical issues involving drug, hormone and antibiotic residues, and significant pathogen levels, and, of course, certainly, regulatory and social perception issues.

Technical Innovation

There may be some doubt that micro-organisms in AD (especially mesophilic at around 38°C, or even thermophilic at around 56°C) would adequately manage the drug and anti-biotic residues, and also likely to be limited in pathogen control. What knowledge we do have indicates that these residues do get through the sewage treatment system in amounts that are small but do affect the environment (e.g. as mentioned above, it has been known for many years that male fish sometimes develop ovaries, from residues of the contraceptive pill, and, more recently, wales getting cancer). More recent research indicates that this pollution is much more serious than previously thought.[130]

Alternatively, it is more likely that TAD (thermophilic aerobic digestion) would crack these components. TAD will, if enough organic carbon is present, will go up and beyond normal pasteurising temperatures. Alternatively, heat from a CHP unit (combined heat and power) may be as good or better.

There is some experience with hybrid systems, coupling TAD and AD together but little scientific measurement to define performance. There is a further technology which uses lime or CKD (cement kiln dust, itself a 'controlled waste') to process and dry AD liquor to produce a granule or pellet of organic-based fertiliser. There is reason to believe that it works and may be economic. If that is so, then this makes a hybrid process a great deal safer and a positive result more secure.

Opportunities

i)   Waste recycling avoiding incineration and the associated production of carbon dioxide. A hybrid, or mosaic, of any combination of these technologies (AD, TAD, pasteurisation, plus drying and/or calcium oxide from quick lime or CKD), as a matter of opinion at this point, has a good chance of taking a significant proportion of hospital wastes, saving the disposal cost, and delivering useful amounts of electrical power and/or heat, plus saleable fertiliser (which, in turn, saves imports) – *safely and economically*. Also note that AD plus a CHP delivers electrical power and heat, 24/7 and therefore contributes to base load on the hospital internal grid (and/or the national grid), day in, day out. It also is politically attractive because it is 'green' and contributes to climate change targets. If the methane is used to heat the hospital (rather than produce electricity) it may attract a *Renewable Heat Incentive* grant on every kWh produced.

ii) This route may also be expected to significantly reduce the risk of the sewage system assisting in the development of resistant strains of bacterial disease, and in pharmaceutical residues entering the wider environment.

130.   Richmond, E et al: *A Diverse Suite of Pharmaceuticals Contaminates Stream and Riparian Food Webs, Nature Communications*, November 6th 2018

iii) Carbon dioxide is produced by AD and in burning the methane to heat the hospital. This gas is 'clean' and can be used in a neighbouring glasshouse in CO2 Enrichment growth of crops for consumption in the hospital or sale elsewhere.

The Synergy

These processes will turn 'waste' into a resource and deliver energy for hospitals, nutrients for food production, carbon dioxide enrichment for growing glasshouse crops (or better still 'dark' growing rooms using LEDs to produce salad crops), cleaner rivers and drinking water with less drug residues, and dramatically reduce risks of carry-over of pathogens and drug residues into the wider environment.

Chapter 16: Global Scale – Quantities & Qualities

Composting as an Industrial Process

There is no technical limit on how big a composting operation could be. However, the ultimate size will be dictated by the end market. If the output goes to supply garden centres all over the country, then the process needs to be near the source of waste. Alternatively, the really big end market, big enough to solve the global problems discussed here, is agriculture and forestry. Then there is common sense in two basic principles. Firstly, minimising logistics means that the composting operation near to be near the sources of the wastes it processes, *and* near to the fields on which it will be spread. Secondly, the people who have the skills, most of the machinery and an in-built responsibility which fundamentally exceeds that of any sector including the regulators themselves, is the farming community who are the guardians of the land and live by any mistakes they make. So, proximity operation, scaled to the sources, skills and lad available are the ultimate target.

The micro-organisms in compost and land are different from us, of course. However, in terms of dietary needs, they are remarkably similar. They are different in that they can tackle much bigger molecules than we can. Just as we might put sugar (a 6-Carbon molecule) on our cereals for breakfast, the micro-organisms can tackle the lignin in wood (with an almost unlimited number of carbon atoms in the same molecule) and they never stop eating, 24/7.

So, compost heaps can break down almost any organic molecule – 'organic' meaning molecules based on carbon chains. In practice, this does apply to liquid plastic but solid plastics take too long for most commercial situations. Given a balance of other necessary food molecules, the compost micro-organisms will crack 100% of the carbon chains in the feedstock in time. Furthermore, it does not really matter if there is any residue before spreading to land, provided that the soil is biologically active (meaning that it already has a reasonable amount of organic matter in it).

World Bank figures[131] in 2012 put the global MSW generation levels as "1.3 billion tonnes per year, and are expected to increase to approximately 2.2 billion tonnes per year by 2025. This represents a significant increase in per capita waste generation rates, from 1.2 to 1.42 kg per person per day in the next fifteen years." So, the quantities are large, and that is just MSW – municipal solid waste. What about other urban wastes? What is the content of these wastes?

Fig 16.2 shows some of the variation in the content of MSW in some Arab countries.[132]

---

131. Hoonveg, V & Bhada-Tata, P: *What a Waste – A Global Review of Solid Waste Management, World Bank #15*, March 2012

132. Etriki, J: *Municipal Solid Waste Management and Institutions in Tripoli, Libya: Applying the Environmentally Sound Technologies (ESTs) Concept, University of Hull*, May 2013

Destination of collected wastes in Europe

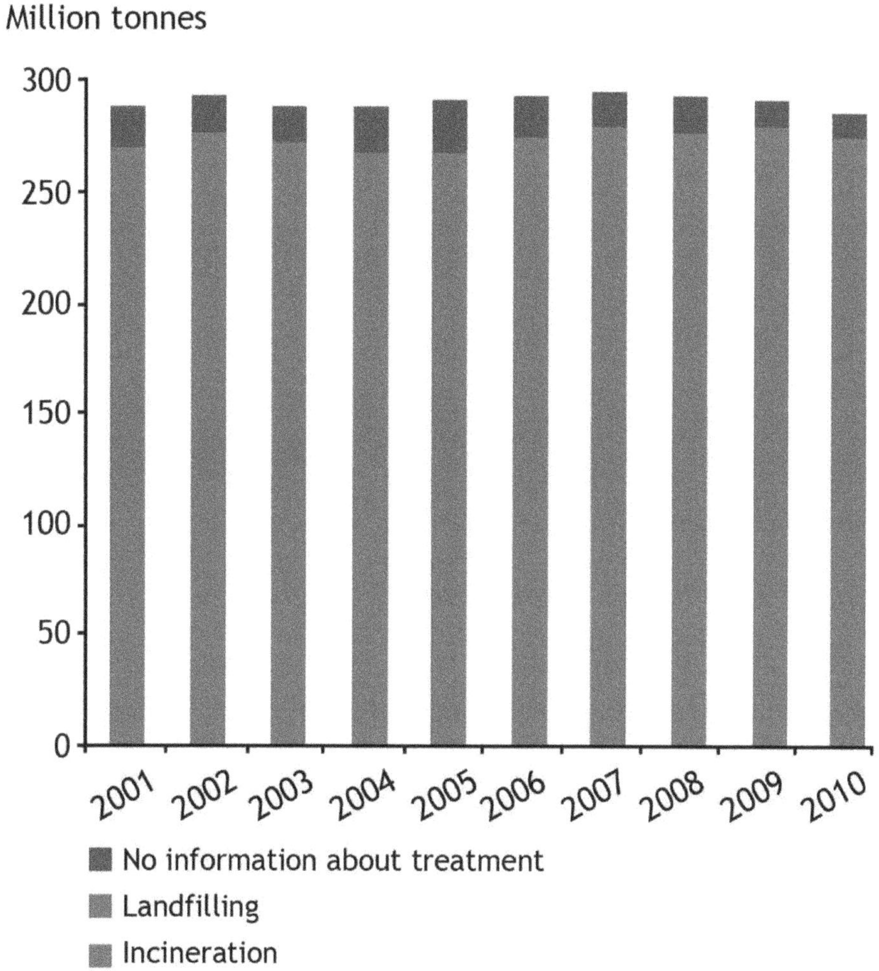
Million tonnes
300
250
200
150
100
50
0
2001
2002
2003
2004
2005
2006
2007
2008
2009
2010
No information about treatment
Landfilling
Incineration
Recycling

Note: The figure covers the EU 27 Member States, Croatia, Iceland, Norway, Switzerland and Turkey

Source: Eurostat, 2012a, 2012c, ETC/SCP, 2013a, 2013b, 2013d, 2013e, 2013f

Fig 16.1: Development of Municipal Waste Management in 32 European Countries, 2001–2010

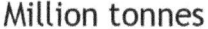

Content of waste in some Arab countries

Country	Organic Materials %	Paper %	Plastic %	Minerals %	Glass %
Egypt	50-60	10-25	3-12	1.5-7.0	1.0-5.0
Jordan	63.0	11.0	16.8	2.1	2.1
Syria	62.0	4.0	7.0	6.0	4.0
Yemen	55.0	14.0	13.0	2.0	1.5
Saudi	37.0	28.5	5.2	8.3	4.6
Kuwait	50.0	20.0	12.6	2.6	3.3
Qatar	45.0	18.0	15.0	4.0	10.0
Bahrain	35.0	28.0	6.0	12.0	5.0

Source: Abou-Elseoud (2008)

Fig 16.2: Municipal Solid Waste Components in Some Arab Countries

Pollution and fertiliser values of wastes in Black Sea states

SOURCES	BOD		N		P	
	t/yr	%	t/yr	%	t/yr	%
Domestic	68,955	6,1	20,294	3,1	6,6655	13,2
Industrial	81,82	7,2	146,934	22,7	2,024	4,0
Riverine	55,318	4,8	49,526	7,7	4,941	9,8
Industrial	934,974	81,9	430,538	66,5	36,876	73,0
TOTAL	1,114,067	100	647,292	100	50,496	100

Fig 16.3: Fertiliser Values in Sewage Effluents in Black Sea States

Note that the content of 'organic' materials is highly variable. So how much is there which could be composted? Frankly, the answers are difficult to be in any way precise about. Maybe over a billion tonnes could be composted annually, globally. As a wild guestimate, maybe enough to fertilise thirty million ha. It could be less and it could be several times more.

The UN complied some data about sewage effluents in the Black Sea states.[133] In 1996 a 'transboundary diagnostic analysis' was carried out of wastes in the Black Sea states. The analysis covered pollution potential in terms of BOD (biological oxygen demand), nitrogen and phosphorous. All of these measures could, alternatively, be seen as of potential fertiliser value.

The point here is that add in global production of sewage and the figure could double the estimate for MSW. Further, adding sewage to separated MSW would enhance the composting process, significantly reduce pathogen risks and improve the fertiliser value of the output. Now add in selected industrial wastes. Probably adding a similar annual tonnage. Bearing in mind the figure from UN sponsored research[134] of typically 21,000 kWh to manufacture and deliver one tone of nitrogen nutrient (fertiliser) these figures indicate the scale of the opportunity.

Now add in hospital wastes (see *Chapter 15*).

Now add in the materials in the *Appendix*. The author, here, has direct experience of composting around five million tonnes of many/most of them. Bearing in mind that a high-yielding arable farm of 400 ha would need around 25,000 tonnes pa of raw wastes, composted, and no other fertilisers, that five million tonnes would fertilise 80,000 ha of high-yielding crops.

Go back to some simple statistics to give an indication of opportunity. OECD (Organisation for Economic Co-operation and Development) figures indicate that one person's output of waste per annum is:

Country	kg per person per annum
USA	750
UK	580
Japan	400
China	120
India	100

Take the smallest figure of waste generated per person – India at 100 kg per person per annum. With 1.3 billion people, that yields 130 million tonnes.

---

133. *Summary of Available Wastewater Data*, *United Nations*, ANNEX 5

134. Gellings, CW & Parmenter, EC: *Energy Efficiency in Fertiliser Production and Use*, *Efficient Use and Conservation of Energy Volume 1*, *UNESCO*, 2004

These figures give some idea as to the opportunity to use wastes for land reclamation, for making fertilisers and growing crops.

It is not just the energy-saving value of these wastes when used as fertiliser, it is also the carbon dioxide saving. One study[135] indicated that the global typical production of carbon dioxide from coal-generated electricity would be in the region of 1000 g per kWh. If the above estimates of wastes which could be recycled to farm land were somewhere near the truth, then it might save the production of hundreds of million tonnes of carbon dioxide per annum, from the saving of fertilis er manufacture alone.

The nature of government is to provide statistics. Frankly, we do not have, nor will ever have, any reasonable level of accuracy about these figures. To collect accurate figures would take a lifetime and be out of date and inaccurate before they were assembled. What is clear from this brief look is that there is an opportunity to do something about global warming with a much more active, enabling attitude to waste as being a valuable resource, rather than using the word 'waste' as meaning a loss. So, the rest of this text is about the technology of managing waste as a resource.

To arrest and reverse global warming, many things have to be done and done quickly. This is just one of the opportunities.

Remediation

Composting is the standard technique for in-place remediation of heavy hydrocarbon contamination of soils.[136,137] If the process is sophisticated enough (as it is) to break long carbon chains which are cross-linked and very stable (such as lignin in oak wood), then it can easily break comparatively short molecules that are found, for example, in petroleum or in a wide range of contaminants from industrial processes.

The only provisions to this are that the material must be spread out far enough and enough time given for the composting process to work. The compost process must have, of course, the basics available involving food, moisture, oxygen and the micro-organisms. Mostly, 'seeding' with a culture of 'special' micro-organisms is not necessary; this earth is well enough populated with micro-organisms, wherever it might be.

135. *Carbon Footprint of Electricity Generation*, *Houses of Parliament*, London, Post Note Update #383, June 2011

136. Butterworth, B: *Reversing Global Warming for Profit*, *MX Publishing*, London, 2009

137. Butterworth, B: *How to Make On-Farm Composting Work*, *MX Publishing*, London, 2009

Chapter 17: Global Scale – Organisation & Reverse Franchising

Land Network as an organisation was conceived in the UK back in 1992 when a farm contractor, Tony Birchnall, went to the Marketing Initiative of the DTI (Department of Trade and Industry of Her Majesty's government in the UK), and said, "I see the writing on the wall for my business, I'd like to look at waste recycling. Can you help?" The project arrived on the desk of the author of this text. What emerged as a project was staggering in its clarity:

How could the resources which already existed in the rural economy be harnessed to recycle to land safely and economically and so as to avoid the purchases of imported mineral fertiliser?

There were two important parameters required to build an organisation:

- a technology and legally-based discipline which earned the respect of all the stakeholders, and
- a recognisable respect for the sovereignty of individual farmers.

In both of these, the word 'respect' occurs. It might also be the word used in terms of respecting nature and its processes in order to create safety and sustainability.

It is sometimes said that no-one has ever made British farmers co-operate on a production operation. Whether that be true or not, the difficulty would be recognised world-wide. The defence-beyond-reason of sovereignty of landownership or occupancy is instinctive in almost all farmers, everywhere (it is pretty common in the rest of the human race, too). Therefore, in this project, it was necessary to avoid any question of compromising the independence of the individual. Instead, something had to be added and obviously so. The model came from JC Bamford, the man who developed what might technically be referred to as a back-hoe excavator but is commonly known, worldwide, as a *JCB*. In order to sell his machines, JCB, the man, gave machinery dealers a 'franchise' but insisted they set up not just a separate department but a separate company and that company would include in its name the letters JCB. He also took a 10% share in that company and sat on their board. So, he knew what was going on.

The farmers' consortium, Land Network, was set up in a similar way but with a twist. The farmers were offered two important safeguards. Firstly, initially, none of the trading was done through the farm accounts. All of it was initially done through the central support operation which was set up as a separate company. So, if anything ever went wrong, it did not affect the farm itself. The central company had no assets and took legal responsibility for compliance with the law and on health and safety matters. That meant, for example, that if what became known as *Central Support*, led by the General Secretary of Land Network, wrote a *code of practice*, and the farm stayed inside that code and there was an accident with injury to a human, it was the General Secretary, not the farmer, who was morally and legally responsible and risked going to prison under health and safety regulations. These safety mechanisms were fundamental. Indeed, the whole ethic of the organisation was, and remained, risk identification, evaluation, isolation, management and all those functions which go into risk analysis and management. For that risk management to be successful in the short and long terms, there clearly had to be a discipline which was imposable and sustainable. In many circumstances, risk management can be imposed and, with willing participants, can be continually developed. That continuation necessarily needs that willingness.

What was the twist to add to the JCB model? Well, farmers could progress to have a share in the organisation and they could actually own it, all of it. The design was complete. It did provide a start which was acceptable to some, even many. As individual farm operations grow, they get to own their own 90% of their own Land Network company with the Central Support unit in the centre owning the remaining 10%. This structure is a franchise where the franchisees get to own the franchisor in what became known as a 'reverse franchise'.

This approach to community-based organisation could be applied anywhere in the world, not just with farmers but also with any trade – plumbers, bricklayers, cleaners, accountants and anything in suits. It takes nothing from the individual except the compliance with a discipline which is, in any case, a necessity in modern society. In return, it gives a support which could not be afforded by a small individual operator and a brand image which can develop into something with significant economic and political clout.

As a national, farmer-owned consortium, the function of Central Support in Land Network was operated by a separate management company which started out as what many would call a consultancy company; the one which worked for the DTI at the start of the project. That company was also structured so that it was progressively owned by its staff. Central Support provided codes of practice based on common sense, farmer-members' own knowledge, technology, regulation and, perhaps most important of all, the developing pool of experience which the organisation had as a whole, i.e. all of its members. The central operation also provided supervision, advice, commercial assistance and whatever support members needed to develop their businesses.

Initially, the farms did all the physical work and Central Support did virtually all the paperwork. The farms were supplied with the necessary paperwork, including data sheets for recording waste 'in' (weighbridge ticket or other means of estimating load quantity and identity) and support to manage their own bank account. Central Support collected the gate fees and took an agreed, small percentage, before passing on the majority of the cash immediately on receipt from the waste hauliers. If and when an individual farm grew big enough and if all the conditions are right for sustaining that growth, then they were set up with their own Land Network company which carried a regional label. The following figure shows the basic structure of Land Network as a reverse franchise. The farms remained as individual, sovereign businesses but they may have operated in a group in order to serve one large supply contract.

Central Support was responsible for disciplining the supply chain which started with 'waste' at the point of its creation and ended with the production of crop and animal products. Increasingly, using the land as a carbon sink to lock up carbon dioxide, was seen as a product.

THE MANAGED LANDBANK USING REVERSE FRANCHISING

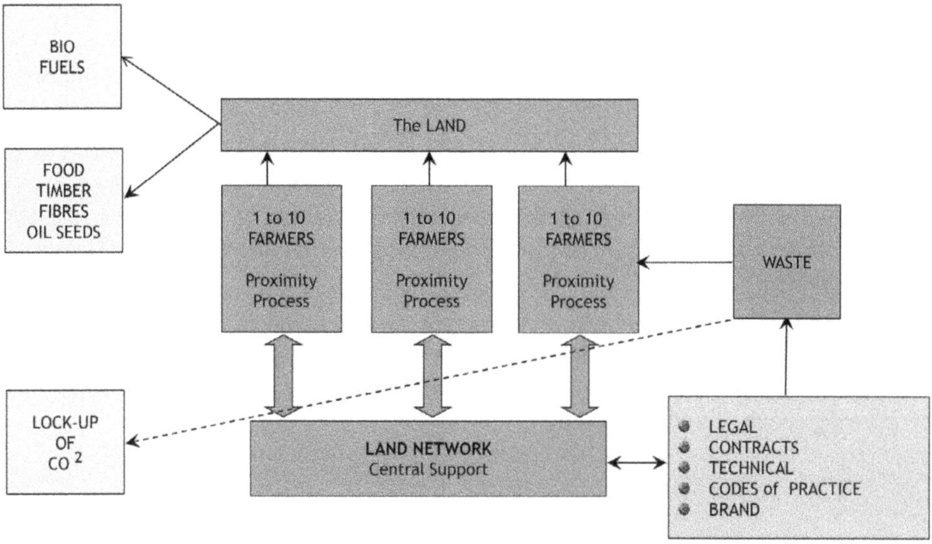

Fig 17.1:The Managed Landbank Consortium Using Reversed Franchising

Over time, some of the sites became industrial scale, commonly handling 25,000 tonnes per annum, and some over 100,000 tonnes. This is a development related to investment in much larger areas of concrete with significant investment in processing facilities. The structure is the classic upside down pyramid, starting with the Central Support directors who support the regional companies. They then, in turn, operate their own industrial composting facility and their own landbank. This structure was developed from the original basic Land Network reverse franchise (it is in fact almost identical) but, in this case, it was designed to develop major industrial processing facilities, capable of composting in-vessel and in tonnages ranging from 25,000 tonnes per annum up to 125,000 tonnes pa in one site.

What this form of reverse franchising does is to harness the available skills and land in order to recycle. It can be built quickly and can be a disciplined but innovative organisation.

Chapter 18: Sub-Surface Reservoirs

This is one of the shortest chapters in this book, but also perhaps the most important. It stresses just one important concept; the idea of a bio-active global carbon sink.

There is a fundamental problem for biological activity in the top few millimetres of soil; it is a very inhospitable place. At times, the radiation from the sun can be dangerous to life. At times, it can be wet enough to drown life. Sometimes, it is just right to plant a seed and those conditions may last just long enough to get the first root down into a more hospitable climate,

There is always a risk to life in that surface layer. However, it is possible to plan what happens just below the surface and create and manipulate a much more stable environment. Fig 18.1 shows these two layers and indicates the management opportunity for growing crops and storing carbon.

TOP SOIL RESERVOIRS

SUBMERGED BED RESERVOIRS

Options
1. Plant seeds and irrigate
2. Transplant seedlings through layer

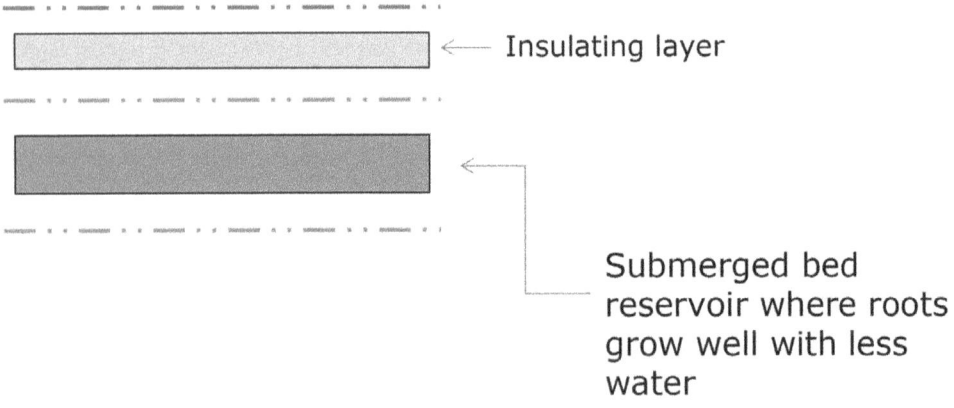

Insulating layer

Submerged bed reservoir where roots grow well with less water

Fig 18.1: Storing Carbon

The existing black soils in the world are known to be both good for growing crops and safe in terms of nutrients leaching into groundwater... so, make some more. A lot more. Soils containing mineral particles, fibres and organic carbon allow better water holding and movement, allow better gas exchange (air and oxygen in and air and carbon dioxide out), are easier to cultivate, and grow better higher-yielding crops, with less disease needing less crop protection chemicals.

The logic, evidence and experience is that high-organic-Carbon soils are environmentally safe. The evidence is that well-composted wastes as the source of organic carbon are similarly safe. We can store unlimited amounts of organic carbon this way. It will oxidise but very slowly if it is below the surface, especially if direct drilling (or zero till) is used to plant crops.

Detail of how to build these sub-surface reservoirs is discussed in *Chapter 21*.

Section 5

Maintaining the Global Bio-Active Carbon Sink

Chapter 19: Sustainability Notes

Sustainability

Sustainability is the key word; in soils, in cropping and in human life.

In tropical areas, man grew up with a *slash and burn* operation in the forest. Trees were cut down, burned and the ash sustained crops for a few years and then the people moved on and repeated the process on another spot. In some areas, the same practice is still in use today. The reason for moving on is that the soils, on their own, are often not very fertile and the rainfall such that nutrients are easily washed away into the groundwater. The fertility is locked up in trees, plants and the detritus (organic wastes from falling vegetation) on the forest floor.

Technology

The farming technology of the last half of the 19th century did feed a lot more people but, in part, is not sustainable in terms of energy inputs and chemical use. Intensive farming in developed countries has exploited the land in a similar way and with similar consequences. Use of high-powered cultivation tools and high inputs of soluble mineral fertilisers have allowed organic matter levels to fall. Cultivation assists the soil micro-organisms to oxidise the large carbon molecules to produce carbon dioxide. This reduction in the large carbon molecules results in changes in the physical strength of the soil structure. As that happens, there is a need for higher power inputs to cultivate the soil; there is a spiral of inputs and the soil is only kept productive by increasing inputs – all of which are energy demanding.[138,139]

Cation Exchange & Base Saturation

Back in the 1950s, Dr. William Albrecht[140] began to look at soils which had been, within living memory, broken out of prairie. These soils had initially yielded crops well but had gradually declined. Whatever farmers did in terms of addition of fertilisers, they could not get back to where they started in terms of crop yield. Albrecht developed chemical models of the prairie soils and the depleted soils and then investigated practical ways of applying available materials to move the depleted soil back to prairie status by using the models as a guide. It worked and still does with Neal Kinsey widening its scope and application.[141]

---

138. Gordon Spoor was a lecturer and researcher at the National College of Agriculture Engineering, Silsoe, UK, for more than twenty years in the 1980s and 1990s. He was acknowledged as a world expert on soil strengths and published many papers on the subject

139. *Understanding Rural Land Use*, *Environment Agency*, NE-1/100-3.SK-C-BEKC

140. There are many references in many places including on the web, for example at Washington State University, Cornell University and others

141. Kinsey, N: *Hands on Agronomy*, *Acres USA*, 1999

Think of feeding the soil and the soil feeding the plant. Now think of the soil as equivalent to a cow's rumen with the alimentary tract as the soil micro-organisms. The soil has enormous numbers of micro-organisms. Neal Kinsey soils reports[142] estimates of the number of micro-organisms in one acre of soil are equivalent in weight to one whole cow. Maybe, according to some thinking, the soil mycorrhiza are the last link in the chain leading into the plant root hair. "It is possible to grow plants on sterile sand with purely mineral fertilisers but generally it does not work so well."

In a study tour shadowing the now world-renowned soils expert, Neal Kinsey who took over Albrech's work, I was shown a farm where Kinsey had been called in to look at a dairy farm where the grass yields had declined and whatever NPK fertilisers (nitrogen, phosphorus and potassium) the farm added, the decline progressed. Kinsey modelled the soil and recommended additions of other plant nutrients, including what farmers call 'trace elements'. It took years but the grass yields did return to 'prairie days' fertility. Something else happened; cows 'took to the bull' first time, had more live calves and the calves grew faster. The cows had less disease, stayed fertile for more years and lived longer. The situation was complex and, from this one example, it would be dangerous to conclude too much. However, this is not an isolated incident. There is a significant amount of credible evidence that getting the right mineral elements into the ground not only helps the plant (and the crop), but also helps the animal that feeds on those plants. It is entirely logical to progress that thinking to suggest that that food chain does progress right into human health and longevity. There is good evidence to support this.

Figure 12.2 in *Section 4* above shows the basic idea of the soil rumen (i.e. the soil universe being viewed as being similar to the rumen of cattle)[143] Using the addition of ammonium nitrate as an example, it shows how charged mineral molecules, called *ions*, can be held in the soil and 'fed' to the plant and, therefore, the crop. These ions are soluble and move when in solution. In a sand, therefore, they are easily leached away. In a clay some of them, those with a positive electrical charge, are held by the colloidal properties (they act like a sponge, holding ions on their surfaces) of the very small clay particles. The *cations*, such as calcium, magnesium, and ammonium are held on the soil colloids which carry negative charges. Fertiliser anions, such as nitrate, sulphate, chloride, however, go into solution and are easily leached away. We have historically limited our thinking to *colloids* being clay particles but it is clear that organic matter as humus is also colloidal. And it is possible to improve humus by a number of methods, some of which are hundreds of times more effective than others. With the right method and the right circumstances, the colloidal properties of a soil can be altered (as in fig 12.3 in *Section 4* above). In other words, it really is possible to change and significantly improve the fundamental fertility of a soil. By increasing this colloidal 'ion bank', it is possible to increase yields and reduce costs. Increasing the bank is partly putting extra cash in (mineral nutrients – NPK and trace elements) and partly having a bigger bank to hold more cash. The basic principle is to add organic carbon.

---

142. Ibid.

143. Butterworth, B: *Managing the Soil Rumen and Ion Exchange*, **Arable Farming**, September 9 2000

Understanding how to manage this bank now becomes easier. When we measure soil pH, we generally tend to just think of aiming at a neutral soil, avoiding acidity – so lime is added. However, as Kinsey says, "You can have too much of a good thing. If too much calcium is added, then the Ca^+ ions push other ions off the negatively charged colloid sites and 'lock' them up." In effect, they are still there but cannot enter the supply route to the plant root. Crop and soil consultant Peter Wright says: "By just using pH to balance the soil with calcium, we miss the bigger picture in that other cations (magnesium, sodium, trace metal elements), as well as calcium, influence pH. It is possible to have a pH of 7.5 with the soil still requiring calcium. Conversely, a soil can have a pH of 5.8 and have adequate calcium but need magnesium and potash to restore the pH to nearer 6.5 which is where soils function best. This is determined by examining the 'base saturation' of the soil. Maybe 70% of arable soils in the east of the UK may have too much lime."

Most soils need 60–74% of calcium, 8–14% magnesium, 1–3% sodium and 4–12% hydrogen. Exactly how much calcium is needed depends on the soil and many other factors, some of which we understand and some of which we don't.[144]

Reducing Inputs

Reducing cultivation by using 'minimal cultivation' or what is termed *zero till* in the USA or *direct drilling* in the UK, will slow down this oxidation of organic matter[145,146] Research by the author of this text for ICI Plant Protection in the late 1980s and early 90s indicated that oxidation of the organic matter occurred from something in the region of 35% per annum on a declining basis with conventional high-input cultivations to around less than 10% with direct drilling. 'Declining basis' is comparable to *half-life* in radioactivity decline. If the figure for carbon release is 35% this year, it will be 35% of what's left next year if the cultivation regime remains the same, and so on.

As all developed societies produce 'wastes', there is a logic in composting what is useful and safe and putting it onto the land to replenish organic matter levels. There is here, however, a potential problem. If the ground is not productively used for crop growth, there is economic loss (the area is not producing anything) and the ground is at higher risk or erosion from wind and rainfall. If it is used for crop growth, it is certainly a potential difficulty to apply compost or wastes to that ground. Composts and wastes tend to be high volume/low value and, therefore, the tonnage needed per hectare may be enough to smother or reduce the growth of the crop. So, there needs to be care in these applications, either in appropriate preparation and application rates or by spreading in 'crop windows', i.e. between one crop harvest and the planting of the next.

Research and experience in the UK is identifying significant advantages in recycling to land. The addition of large quantities of organic wastes or composts has a number of effects on soils. Application rates in the range of ten to twenty-five tonnes per

144. Kinsey, N: *Hands on Agronomy*, *Acres USA*, 1999

145. Gordon Spoor was a lecturer and researcher at the National College of Agriculture Engineering, Silsoe, UK, for more than twenty years in the 1980s and 1990s. He was acknowledged as a world expert on soil strengths and published many papers on the subject

146. Butterworth, B: *How to Make On-Farm Composting Work*, *MX Publishing*, London, 2009

hectare are allowed by the regulators and do produce significant physical and biological responses which have economic and environmental advantages. However, there is no scientific reason for limiting the amount of well composted wastes applied per ha. Organic material (in large quantities) will hold moisture, release nitrogen slowly, change the fungal population of the soil, reduce crop disease and herbicide use, and reduce power used in cultivation.

Soil Mycorrhiza, Crop Diseases and the Closed Loop

Some years ago there was a breakthrough in identifying soil 'glue' and its relationship to soil crumb structure, soil tilth, gas exchange, soil water movement and retention, crop stress and disease.

Sara Wright,[147] a microbiologist researcher at the Soil Microbial Systems Laboratory, Beltsville, United States Department of Agriculture (USDA), was part of the team which identified and named *glomalin* as the protein which appeared to be the binding agent in the formation of soil aggregates and, therefore, the controller of so many practical functions in which the operator in the field is so interested. This breakthrough can be coupled to advancing knowledge on mycorrhiza and can open up the route to lower costs in cultivations, lower costs in crop protection and higher crop yields. The team found that tillage tends to lower glomalin levels.

Soil aggregation
During the 1980s, quite a bit of academic research was done on soil aggregation and the formation of crumb structure. It was clear then that cultivation increased the mineralisation of organic matter which was closely related to the formation of aggregates. In turn, it was also clear that such loss of aggregation also resulted in reduced gas exchange, reduced water movement, reduced crop growth and increased power requirement for cultivations. As many arable farmers know to their cost, the pressure to get on and do the job gets the operation onto a descending spiral of declining soil structure, pressure on yields and increased costs.

Research also confirmed that what grandfather knew was indeed true and now could be measured in scientific terms; i.e. that some crops were more capable than others of putting organic matter back into the soil. Hence the historical practice and interest in 'green manuring' (growing a green-leaved crop and ploughing it in) and, more recently, ploughing unwanted cereal straw back in. The research led to the *hierarchical theory* which proposed that soil stability depended on macro aggregates which were based on micro aggregates held together with organic matter such as pieces of roots, stems, leaves and other plant remains. Hence, the interest in cultivations which resulted in the breakdown of organic matter. In turn, the micro aggregates were held together by small, negatively charged clay particles. That was clear enough but it has become progressively apparent that the binding of micro aggregates into macro aggregates is not only dependent on active plant roots and decaying plant remains, the presence of fungal hyphae is also a fundamentally important part of this activity. The identified fungi are *vesicular arbuscular mycorrhiza* – VAM fungi for short. There is no doubt that

---

147. Sara F Wright & Kristine A Nichols are with the SDAARS. This research is part of *Soil Resource Management* and *ARS National Program* (#202)

the biological activity of mycorrhiza is of fundamental significance in terms of science, technology and commercial farming[148,149]

The function of fibres in the soil is interesting; in a peat there are fibres of hemicellulose and lignin which may be hundreds or even thousands of years old. These clearly have a physical value in gas and water management when mixed into soils. There is some logic and practical evidence that synthetic fibres, such as nylon and polypropylene from carpets, can and do have a similar function and value.

Active management of soil mycorrhiza can lead to better crop protection for less money, with a bonus of better yields. Research into the function of mycorrhiza, the fungi which surround plants, indicates that we can develop a line of thought in methods of field procedure to get more control for less cash.

Mycorrhiza are a group of fungi which are closely associated with plant roots. In practice, in the field, that means all plant roots, all of the time. The relationship is symbiotic; both parties get advantage. We know that the relationship is very complex and that it is more active at some times than others. As is always the case, it is not possible to manage with any consistency unless that management is based on at least some knowledge of the mechanisms involved in the process to be managed. We think, from the research so far, that we could, at least in part, make significant progress in managing the relationship. The advantages may be very substantial because the mycorrhiza dictate the plant's relationship with the soil. In effect, they have a very major role to play in water uptake, nutrient uptake, and the whole of the metabolism of the root and, therefore, the plant and the crop as a whole. This will have significant effects on how the plant (and, indeed, the crop) competes with weeds and pests. This, then, is fundamental to crop husbandry and commercial success.

A researcher called Rejon, working for the USDA, has worked with a team on mixed plantings of wheat and perennial ryegrass.[150] A trial was carried out on potted plants in the lab under controlled conditions. Selective herbicide was applied to remove the ryegrass. It is possible to produce plants without mycorrhiza on the roots and, in this trial, control was significantly better when mycorrhiza were present. Yield was also higher. The researchers concluded that the mycorrhiza have fungal hyphae which are associated with several plants and, when the weed crop is weakened by the herbicide, these hyphae may be associated with moving nutrients from the region of the weakened plant roots, maybe even from inside the root, to the stronger and un-weakened crop plant.

148. Berg, B et al: *Plant Litter, Decomposition, Humus Formation, Carbon Sequestration, Springer*, 2003. This is one example of many hundreds of references to organic matter degradation and formation of humus in the academic literature and on the web

149. Harrison, MJ: *Signalling in the Arbuscular Mycorrhizal Symbiosis, Annual Review of Microbiology*, 2005 #59 p19

150. Rejon, A et al: *Mycorrhizal Fungi Influence Competition in a Wheat-Ryegrass Association Treated with Herbicide Diclifop, Applied Ecology*, 1997 #7, pp51–57

The questions raised by this research are many. Firstly, it may well be (even likely) that some mycorrhiza species are better than others at helping the effect of herbicides or other crop protection chemicals. Maybe some are better with certain crop weed combinations. Maybe some react better with some crop protection chemicals than others. Turning the questions round, maybe this could explain why some chemicals appear to work better in some situations than others i.e. the right mycorrhiza have to be present. Determining all the correct factors is very complicated but we do have some good leads.

Potatoes & Cereals
EcoSci, based in Exeter, some years ago, did quite a useful amount of quality research on the disease control effects of composts made from different materials. Some of these figures are shown in tables one & two below from an extract from the paper in *Resource*[151] (the journal of the American Society of Agricultural and Biological Engineers). Clearly, the mature greenwaste compost had something special. Another laboratory trial by EcoSci gave 100% control of brown rot in potatoes grown on organic composts. While, in these trials, there was no examination of the species of mycorrhiza present, the implication could be that the different sources of material and their different conditions affected the activity of the mycorrhiza and how they helped the plant overcome disease. The conclusion that managing the soil organic matter in order to manage the control of diseases better does, in the light of long term, practical experience, look attractive. It must, at least in part, be involved in unravelling the story.

Crop Stress
Research on mycorrhiza shows that they become more effective in assisting their plant hosts as stress levels rise. If water and nutrient levels are low, then plants grown under conditions sterilised of mycorrhiza are significantly worse off and this effect tends to be progressive; as the stress increases, so does the effect of not having associated mycorrhiza. Again, turning the question round, this is likely to be part of the explanation of why sometimes a crop under stress conditions does better than another for no apparent (previously known) reason.

There is no doubt that mycorrhiza are living organisms and that living organisms need food. Mycorrhiza feed on organic matter. Arable soils with low organic matter can yield well but the technology has to be poured in at ever-increasing levels; mineral fertilisers, crop protection chemicals and mechanical power. Over time, the system progresses closer and closer to hydroponic systems. These systems can work exceptionally well and, there is no doubt, many millions would have starved, world-wide, if these systems had not been developed. However, crop production using high levels of organic matter does have major advantages, especially as costs rise for petroleum-based fertilisers, crop protection chemicals and mechanical power. The best technical results is most likely to come from using high levels of organic matter in the soil *and* high-tech inputs of agrochemicals and power. The two are not exclusive of the other.

---

151. Butterworth, B: *Clamping Down on Compost, Resource, American Society of Agricultural and Biological Engineers*, April 2006

Plant Metabolism

These VAM fungi are not just involved in the physical characteristics of the soil, they affect, are directly involved with, and are likely to be the major route, for the nutrient uptake by the plant.[152,153] There is a very close relationship between VAM hyphae and plant roots. Normally, all plant roots are covered in VAM as symbiotic assistants. They (VAM or roots) are known to be deeply involved in the uptake of phosphate but there is evidence that they also increase the absorption of sulphur, magnesium, iron, zinc, copper, manganese and probably most other things which the plant needs.

These mycorrhiza are also involved in anti-biotic and pro-biotic activity.[154,155,156] There is no doubt that soil mycorrhiza are an effective and practical tool in the management of soil structure in the field. So, management of these VAM fungi is a biological technique which can lead to greater financial margins. There are likely to be many factors involved in managing a very complex population which we can't see and of which we have, as yet, limited knowledge. However, there are two very clear factors we can start with and they are closely associated; organic matter and cultivations.

Soil Organic Matter

Normally, around half of a plant's weight is below the surface. So there is plenty of scope to accept that there is at least a significant amount of organic matter in the system. We also know that cultivations oxidise organic matter. So, if the process must cultivate there follows two logical rules; firstly cut cultivations to what is necessary – and only that. Secondly, relate organic matter additions to the violence of the cultivations; if you hit the soil harder with multiple passes of power driven equipment, then higher levels of organic matter additions will be required to replace that which has been oxidised away. *Recreational tillage* (i.e. that which is not really necessary) has an expensive consequence.

Where will arable farmers get the organic matter from? Certainly, conservation of what is produced in the field is the starting point. However, unless the ground is direct drilled or there is some form of significantly reduced tillage, imports will be necessary and an obvious source is waste from municipal authorities and industry. Millions of tonnes of organic wastes are landfilled every year at enormous cost to the rate-payer and the environment. Farming could close the recycling route and give sustainable operation to some of those other businesses. As Dr. Tim Evans, then of Terra Ecosystems (the company which handles London's sewage products), said[157] "We find that many of our customers using biosolids (processed products originating from sewage) against a planned programme report a reduction of crop disease from the start and it builds up."

152. Kinsey, N: *Hands on Agronomy*, *Acres USA*, 1999

153. Rejon, A et al: *Mycorrhizal Fungi Influence Competition in a Wheat-Ryegrass Association Treated with Herbicide Diclifop*, *Applied Ecology*, 1997 #7, pp51–57

154. Butterworth, B: *Reversing Global Warming*, *Refocus*, September/October 2006

155. Harrison, MJ: *Signalling in the Arbuscular Mycorrhizal Symbiosis*, *Annual Review of Microbiology*, 2005 #59 p19

156. Butterworth, B: *A Top Idea That Holds Water*, *Water and Effluent Treatment News*, October 6th 1999

157. Evans, T: Personal communications with the author

We now know that biosolids, and the new range of products beginning to come from them, have a particular 'inoculating' effect with soil micro-organisms and this is another area of soil management which we can begin to manage better.

The pieces of the jigsaw are beginning to come together. We have paid little attention to soil biology in the last twenty, even forty years. Maybe, now we know a bit more about aggregation and how it is related to soil micro-organisms and crop disease, we know enough to make a start on managing that biological balance to margin advantage.

Back in the early 1990s, EcoSci was a research company in the UK involved in compost research. Table 19.1 shows reduction of crop diseases in plants grown on 100% compost in the laboratory. Such measurements are possible in the laboratory but difficult in the field. However, widespread observations of conditions on farms supplied by the water companies which are responsible for recycling sewage to land confirm that there is a disease control benefit. There appear to be two routes of action. Firstly, dressings of mineral fertiliser such as ammonium nitrate tend to produce flushes of growth which is soft, fleshy and, therefore, more subject to aphid and fungal attack. It also appears that only partly decomposed composts encourage pin moulds (penicillins) and discourage many other organisms. While at least partly understood, it does appear that there are both pro-biotic and anti-biotic effects of using the soil itself as the final stage of the composting process i.e. adding the compost before it is too well rotted (in the original research, there was no statement in this paper of why the crop disease was reduced or totally controlled. At the time, the reason was either not known or not clear. As this text shows, we do now know).

The reduction of drought stress can be major. As an example, again back in the 1990s, Alwyn Moss was a contractor based at Mildenhall in Suffolk, UK, on the edge of the American Air Force base. The local soil is a blowing sand with frequently less than 380 mm of rainfall per annum – the UN definition of desert. Moss took shredded newsprint used as bedding in the racing stables in Newmarket and composted it before adding it to the land at the rate of up to 240 tonnes to the hectare.[158] That newsprint compost will absorb between six and ten times its own weight of water. So, ploughing in the material in the autumn before planting fodder beet in the following spring means that it is possible to grow a useful crop without the irrigation needed by his neighbours. A wider survey of farms in the same area showed a 45% increase in sugar beet yields and an 80% reduction in fertiliser costs for farms using heavy dressings of sewage products when compared with neighbours.

Reductions in the energy used in cultivations following additions of wastes to land are difficult to quantify. Conditions in the field are so variable that scientific statements are clouded or imprecise. However, most workers agree that there is an effect which is, of course, greater where the soil organic matter was previously depleted. Some researchers will offer subjective agreement that these effects may be over 10% reduction in such cases and may be much more significant in seedbed preparation. Land Research has experience of over 50% saving in cultivation energy on heavy land which was ploughed every year and had been in a system with heavy dressings of compost for at least five years.

---

158. Butterworth, B: *A Top Idea That Holds Water*, *Water and Effluent Treatment News*, October 6th 1999

When the value of plant nutrients *(table 19.2)* are brought into the equation, then one is left wondering why farming is not 'farming' waste first and producing crops as a by-product side line.

One of the reasons for slow adoption of waste into farming as part of the system has been easy, 'clean' alternatives and not enough economic pressure. Another may have been direct concern about disease and this is certainly likely to be a problem with supermarkets who may wish to control the image of their inputs. There are two developments in the UK which may be of significant value here; *deep clamp* composting and thermophilic digestion (TAD).

Suppression of Crop Diseases

Soil organic matter and mycorrhizae are closely bound up together as bio-active carbon. The organic carbon is the food for the fungi. The fungi have a very significant function not only in feeding the crop but also in the dynamic balance between the crop and crop disease.

Table 19.1: Suppression of Crop Diseases by Recycled Organic Material (ROM) Composts

Disease	Very Mature	Mature	Paper Waste	Other
Foot rot of cereals	27%	28%		
Brown foot rot of cereals	62%	86%	37%	
Take-all of cereals	46%	81%	66%	
Blight of peas	NS	66%	61%	Sewage sludge and green waste 49%
Clubroot of brassicas	100%	100%	100%	
Black scurf and stem canker of potatoes				Spent mushroom compost and green waste 49%
White rot of onions	75%	90%	67%	

Fertiliser Value of Composted and Digested Wastes

Guideline figures only. Wastes are highly variable and these figures must be taken as useful guides only, to be modified according to specific circumstances.

Table 19.2: Nutrient Figures in kg plant nutrients per tonne of dry material

Original Source	% dry matter	N	P as P205	K as K20	Others
Green waste	40 – 70%	10	10	10	Useful trace elements
Sewage – not digested	4%	50	60	2.5	Useful trace elements
Sewage digested	4%	40	60	2.5	Possible heavy metal pollution
Separated garbage (MSW) 23% biodegradable	30 to 60%	60	90	40	Very variable
Liquor from co-digested sewage and garbage (MSW)	10%	50	70	35	Heavy metals very unlikely

There are, of course, many possible sources of suitable organic matter. The best is likely to be the roots of a healthy crop. Such roots will be at least half the total weight of the whole plants and we know that they have the right mycorrhiza attached. Composts made from municipal wastes, paper and other industrial waste will provide good source material provided appropriate codes of practice are followed. One of the best imported sources of organic material will be biosolids. On that subject, we must move away from 'sewage' because the unprocessed material has begun to be associated with the possible perception of risk. Modern UK sewage derived products are very underestimated materials and significantly safer, in some cases, than mineral fertilisers which rarely are given a second thought on safety issues. However, remember that digested sludges are different and that thermophilically digested sludge is pasteurised and that puts it on a safety level with milk. The biggest concern about biosolids expressed by regulators is the heavy metal content. One of these, to give most concern, is cadmium. There is more cadmium in many mineral fertilisers than in a modern UK sewage sludge. Why press the subject of biosolids? Well, it does look as if the bacteria active in the digestion process just might be part of the story on managing soil biology better. Some of the results with biosolids do appear to be associated with mycorrhiza, the right ones, and crop disease control for less chemical. Whatever the original source of organic matter, there is well-established evidence of effect on crop disease.

Suppression Of Soil-Borne Fungus Diseases

Table 19.3: Using Crops Grown in Lab in Different Composts; % Shows Level of Control

Disease	Compost Made From		
	Very Mature Greenwaste	*Mature Greenwaste*	*Paperwaste + Greenwaste*
Brown foot rot of cereals	62.3%	86.5%	65.7%
Take-all of cereals	46.5%	80.6%	65.7%

*Source – OFTec & Bill Butterworth

Chapter 20: Farming and Forestry for 2030

Composting Wastes

The basic process of composting, or any other aerobic biological process, involves four needs; a feedstock with reasonably balanced nutrients, an appropriate population of micro-organism species, moisture (water), and oxygen. Generally speaking, if the material is of plant or animal tissues, it will have enough nutrients to react reasonably naturally in a compost process. 'Seeding' of an outdoor composting process with a culture of micro-organisms or some active compost from the previous batch is not normally necessary. In most environments, there are plenty of micro-organisms and of a wide enough range of species to operate the process. If there is not enough oxygen, the process will go anaerobic, become odorous and slow down. If there is not enough water, the material may well go dark in colour but the process will be incomplete and will start up again when the material is incorporated into the soil after spreading.[159] This may not matter unless there is a shortage of nitrogen and, in such a case, the micro-organisms in the soil may preferentially use the easily available nitrogen in the soil to build their own bodies in order to attack the energy source (the carbon) in the added compost. Farmers sometimes call this *Nitrogen starvation*. The nitrogen is not lost and will be available to the crop at a later date when the biological activity of the soil catches up with balancing the nitrogen status of different fractions of the soil. This was described in a composite form in *The Straw Manual*.[160] Previous to this date, farmers in the UK faced a ban on burning straw behind the harvester and it was necessary to find out how farmers could incorporate unwanted straw into the soil without high costs and/or loss of yields. That book collated the available research from world-wide sources and showed that the soil will live quite well with what might appear to be enormous imbalances (as with the high carbon content of cereal straw) provided it is given a little help and time. The basic rule when changing to a system which involves putting large amounts of carbon into a soil is to add enough nitrogen fertiliser in the first year to allow the soil micro-organisms to build the protein of their own bodies, allow a little less nitrogen in the second year and less or none by the third or fourth year. After that the soil system will cope because the added nitrogen does not leach out and the soil micro-organism population has changed in species and population numbers to give the biological activity required to deal with the new regime.

A compost process is basically the same. It is worth going back to the basics of composting for a moment; the four needs of feedstock, micro-organisms, moisture and oxygen. It might help to add 'time' at this point. If one of these basics is either not there, or not in the right quantity, then the process will change or slow down. For example, if there is not enough air (oxygen – this is an aerobic process), the process will slow down and either stop or go anaerobic which will give off a bad smell. If there is no change of

---

159. Butterworth, B: *How to Make On-Farm Composting Work*, <u>MX Publishing</u>, London, 2009

160. Butterworth, B: *The Straw Manual, <u>Spon</u>*, 1986

gases, the process will eventually stop. Similarly for water; lack of it will slow the process eventually to the point of cessation. Exactly the same applies to the other two inputs of feedstock (obviously) and micro-organisms (less obvious but quite interesting).

An idealistic view[161] of a composting process graph of temperature against time is shown below. In practice, this will only occur in a compost of uniform material, uniformly shredded and uniformly aerated. Most actual operations will produce a patchier picture although of the same progression.

The first peak in temperature is mainly of bacterial activity. This is the risky time in terms of lack of oxygen and odour production. Deliberate oxygen starvation can and does produce methane gas which can be used in heat and power generation. The top of this curve will normally be targeted at 55 to 75°C. Above this range, the temperature gives rise to increasing losses of nitrogen (which is, of course, valuable as a fertiliser) and increasing risk of fire. The second peak is of fungal activity; mainly *pin moulds*, which are penicillins. The trough in the middle is mainly of actinomycetes, the fungi which give woodland its pleasant smell after rainfall. The curve on the right falls, but never actually to ambient, unless the material, or the ground onto which the compost is spread, is frozen solid.

COMPOST TEMPERATURE CURVE

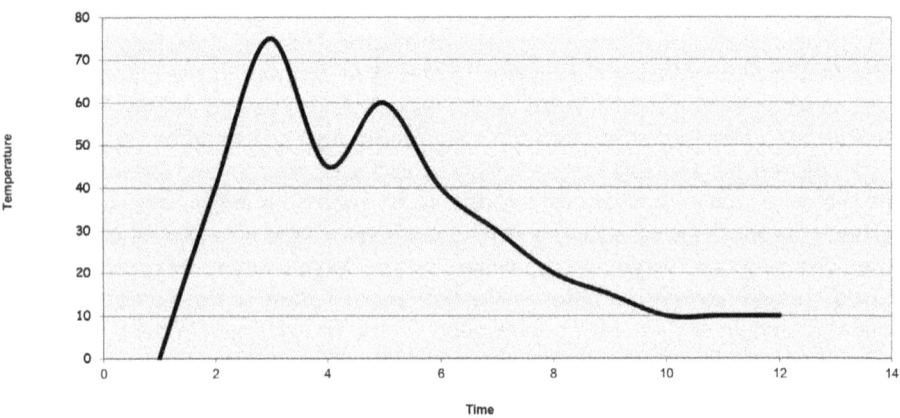

FOUR BASIC PROGRESSORS
1. Live Organisms
2. Food
3. Moisture
4. Oxygen
If temperature is off-curve, one of these four is sub-optimal

Fig 20.1: Compost Temperature Curve

161. Butterworth, B: *How to Make On-Farm Composting Work, MX Publishing,* London, 2009

Following the logic of that curve and the four basic inputs into the process, it becomes clear that it is possible to influence the direction and speed of the process. It is possible to speed the process up and to slow it down but never make it quick; this is a biological process and these, generally speaking, are complex and take time. It is possible to make the process more odorous or less odorous but never completely odourless. It is possible to make the output look dark in colour and friable, but that does not necessarily mean that the process is complete and all the soluble nutrients have been incorporated into the bodies of micro-organisms, turned into humus and made safe to spread without pollution risk.

Composting and Direct Incorporation

In terms of basic technology, if a material is compostable, then it can be put directly onto the soil and incorporated either by machinery or just left on the surface. There are plenty of examples in the Americas, Australia, and all over the world, of 'trash farming' which normally involves leaving previous crop residues on the surface either to help control erosion of just save energy and cost. In the UK, we actually do the same but call it adding a 'mulch'. There is adequate practical knowledge available in the areas where it is practiced, plus capable machinery including grain drills, to plant and establish a following crop successfully. Logically, if that is the case, then it is similarly possible to spread wastes in a similar fashion, left on the surface or with varying degrees of incorporation, *provided* some thought is put into the suitability of the waste in terms of crop nutrients, and toxicity[162] In terms of chemical principle, there is little or nothing that can be done in a compost process that cannot be done in the soil. However, the composting process can reduce or assist in the management of some important aspects of risk.

Firstly, the composting process is at a higher temperature than the activity in soils. This means that chemical, physical and biological risks can be held within the process area, monitored and often speeded up (compared with in-soil processing). That is a significant management advantage in managing those risks. However, in reality, concentration of an activity also brings its own risks and the most obvious and common one in this case is the odour so often associated with badly managed composting sites. However, there is a substantial risk with concrete in that, if and when the concrete is not covered with compost, and if there is heavy rainfall, then there is likely to be very large amounts of collected water. These quantities will get greater as we get into global warming. The regulator's answer will be to have large-enough storage of the run-off water in sealed and 'safe' areas. The concrete and sealing of the run-off lagoon will have their own environmental cost and this is another case of the regulations increasing, not decreasing, environmental cost. Putting liquids onto windrow composting, without run-off, is difficult. However, liquids can safely be put on compost without concrete underneath by spraying controlled amounts on top of a *deep clamp* (usually three metres deep), or a static pile or 'lagooning' on top of a heap (see *Chapter 15*).

Secondly, composting allows a time buffer. Time in industrial terms probably costs money; there will be concrete, machines, energy and manpower involved. However, in farming, the first advantage is in holding the material for a suitable crop window. A further advantage is the time factor, this gives the opportunity for the management of

---

162. *Understanding Rural Land Use*, *Environment Agency*, NE-1/100-3.SK-C-BEKC

processes in a relatively controlled environment and that, in turn, will allow experience to predict the time needed to achieve desired standards and that is helpful in designing standards which are publicly acceptable.

Spreading

There may be some operational risks when spreading materials which have not been composted. Mineral salts or those which may release ions can be tested for conductivity. Examples of these might be sodium chloride (which is commonly used as fertiliser for sugar beet) or acidic industrial wastes such as sodium sulphate (which, again, has fertiliser value). There are also more complex industrial wastes which may contain ionised materials. An example here could be the waste from welding wire manufacture which contains ions of calcium, copper, sulphate and hydroxide (quick lime and copper sulphate are often low in soils in some areas of the UK and applied as fertiliser). Another example is the 'waste' liquor from yeast manufacture where the yeast has been grown on molasses; the residue contains significant amounts of dissolved nitrate. When tested, these will show high conductivity and direct application carries a risk of scorching or scorching the green leaves of plants. There is a more insidious risk in high conductivity materials. They have lower surface tensions and pass through soils more easily and, therefore, increase possible risk of leachate into field drains and groundwater. Where these risks are evaluated as being significant, the alternative of composting becomes more attractive.

Odour

There is one really important potential advantage of direct application: odour control. Where liquid or solid materials have a significant odour risk, direct application to land, by injection, dribble bar, or by spreading to the surface and incorporating by cultivation shortly afterwards, may make direct application significantly attractive.

Effect of Waste to Soil

There is no doubt from the above discussion that organic matter can and does have a major effect on soil mycorrhiza and, through them, very significant and at least partly manageable effects on soils and crops. These effects are all of very significant economic benefit. It is here that 'wastes', especially if they contain significant carbon as large molecules, and preferably a balance of nutrients including all the traces which plant scientists look for, offer a very significant benefit. The condition is that there is a direct relationship between mycorrhiza metabolism and the wastes added. There are, as ever, two components of management which can create that as a successful relationship; the knowledge of the farmer who owns the land and the expert knowledge of the technologist.

Process Capability and Safety

The soil, with the organisms in it, is a staggeringly flexible and tolerant universe. The micro-organisms evolved with plants in an integrated and inter-related package which, on the whole, works rather well. Farmers themselves generally do not go somewhere else at 5pm – if they make a mistake, they know they will suffer loss in the short run and possibly hand it on to their children. Add to that an organised discipline based on fundamental science and applied technology and there is the basis of a sustainable operation. The question then arises as to whether that 'organised discipline' is the farmers and their organisation, or the state. The logical answer is, of course, a partnership of all three. However, in Western democracies, there is a tendency for state regulators to feel safe if they regulate for those who are not responsible, to the detriment of those who are. When responsibility is taken away from people actually on the ground, either by big business or by the state, then the system will break down and the environment will suffer. Farm-based systems do not have that fundamental fault simply because the farmer has to live with his mistakes and pass them on to his family and children.

There is one test which is above all laboratory tests and that is *bio-assay* which, itself, can be done in the lab but is better still if it is *in vivo* i.e. in the real living world. It is not in the nature of a regulated society to just allow operators to just get on with it, so laboratory tests are, and will always be, the basis of scientific appraisal of any situation, including the possible use of wastes on the land to grow crops sustainably. Nevertheless, when all the lab tests are done, there is one useful advantage of composting; it is a buffer in the chain between production of wastes and use on the land. As a guide rule, if a material is put into a compost heap, and the process follows normal patterns of temperature progress, time, odour, and visual appraisal, then it is probably safe. Having said that, even if not demanded by regulation, testing output of the composting process is good scientific procedure and essential for managing sustainability.

As an extension of this thinking, all of the farms in the Land Research's composting network in the UK complied with not only regulation but beyond that using best available technology to evaluate composts before putting to land. Some of these farms were still not able to detect anything but sustainable benefits after over twenty years of putting compost on the same land for that period. Yields were still creeping up, crop disease was slowly creeping down and no one, including the UK Environment Agency (despite exercising their powers very actively) was able to identify negative effects. There is an important principle here; long term monitoring is a fundamental of managing sustainability responsibly.

There does not have to be a choice between biofuels and food. Farming can harvest sunlight to produce energy **and** food. There is no reason why a global effort to produce biofuels cannot be accompanied by a parallel global effort to produce food. Indeed, there is every reason why the two should be progressed together as one package. Here's how.

Cutting down the rainforests to grow crops will not stop unless the crop is not allowed (by financial or political pressure) to be palm oil for biofuel production. The only route on the horizon at present with the remotest chance of stopping rain forest destruction is to allow, or preferably force, the world's top polluters to buy and protect rainforest in an 'offset' scheme. It has been talked about and 'progress has been made' but a *compulsory* scheme is not in place to halt *all* destruction and, frankly, is likely to be a long way off; political wrangling is, as usual, the death of necessary progress.

Case study. There are three generations of one particular farming family farming in the north of Lincolnshire, UK. They have a dairy herd and some beef with 800 acres (about 325 ha) of mixed rotation, mainly cereals and oil seed rape. To give a guide, the farm would generally expect winter wheat to yield, average all fields, 3.5 to four tonnes per acre (eight to ten tonnes per hectare).

The two brothers joined Land Network, the farmer-owned consortium, just after the turn of the century with a small scale composting operation building slowly to a permitted operation of 25,000 tonnes throughput per annum. That throughput is mainly green wastes (from gardens and collected by the local municipal authority), timber wastes (mainly MDF containing ureaformaldehyde as a nitrogen source) from furniture factories and some waste liquids (they handled, or planned to handle, nearly all the materials in the *Appendix*). These wastes received a 'gate fee' in the range of £15 to £35 per tonne delivered (2013 values). The farm was building up to fill its permissions for its 75,000 tonnes per annum recycle-to-land facility.

The farm soils benefited from the muck from the livestock but would normally expect to spend around £50,000 pa on purchased mineral fertiliser, mainly nitrogen. It took six to eight years (depending on the individual fields) to switch from purchased mineral fertilisers to only compost from the wastes. No mineral fertiliser input was achieved (note that British farmers import over £1 billion worth of mineral fertilisers pa).

Both brothers 'do what has to be done' as a team, driving the composting operation and the biofuel operation. From oil seed rape (OSR), they get 1.5 tonnes per acre (over 3.5 tonnes per hectare). One tonne of OSR seed will give 330 litres of oil and that, after process, will give 330 litres of biodiesel and 65 litres of bioglycerol.

The farm provided biodiesel at the 100% level and to the European standard EN14214, for the trials for CNH (*Case New Holland*) in their world-wide trials of their agricultural diesel engines. As a result, all CNH agricultural diesels carried a full warranty for biodiesel up to the 100% level (subject to filter and fuel standard conditions). The farm also supplied biodiesel for similar trials in Volvo truck engines.

The point about this case study is that it demonstrated that food and fuel production can go together and it makes good farming sense to do so, it makes good logistics sense to operate both together on a proximity principle basis, and it makes good financial sense to operate a balanced product and business package.

Phages in Sewage

In the 1990s there was a desire, in what we previously called the USSR, to find the next generation of anti-biotics. The logic was glaringly simple. All species have something preying on them. There is the old adage of "little fleas have smaller fleas on their back to bite 'em... and so on and so on... *ad infinitum*". So, where are there lots of bad guys, because if we can find them, there must logically be good guys preying on them. The obvious place to look, for the researchers at the time, was in sewage. They had plenty of that and it was already there with a 'negative cost'. Sure enough, when they looked, they found human pathogens. Sure enough, when they looked harder, they found the good guys attacking the pathogens. The good guys, anti-biotics, were identified as phages. Phages are just sub-optical for the naked eye but can be easily seen with an optical microscope; they look like multi-legged spiders. This, then, is one of the reasons why sewage products can be beneficial in reducing crop disease. Around this time, I was working for CARE, the international charity, on a project in Georgia on the Black Sea and was fortunate to visit the Phage Institute in Tbilisi. International interest in phages slid for a while but by 2015 there began a real global interest in new anti-biotics and that again has raised the interest in phages. I remain of the view that untreated manures probably contain live phages in many situations and this is likely to be one of the reasons why claims that organic farming results in less crop disease are, at least in some situations, justified.

Economics of Recycling to Farm Land

Until the end of the Second World War, the mass production of low cost mineral fertilisers and the logistics of global distribution limited the impact of these potentially highly productive fertilisers on world food production. In the 1950s, however, the impact was staggering. When mineral fertilisers were low cost, up to around the late 1990s, the developed farming systems of western nations found it convenient, productive and profitable to use them; to the point of not bothering with recycling of wastes. The USA developed the *Land Bank* programme to pay farmers *not* to produce and Europe even invented *set-aside* to limit production. Both of these, to the shame of the countries concerned, occurred when over half of the world was short of food.

The global rises in fossilised fuel energy costs in the 1990s onwards have made that attractiveness less and less profitable. The scope for recycling 'wastes', however, has become dramatic. Proximity recycling gets trucks off the road. In the UK, British farmers in the first two decades of the 2000s were spending around £1 billion pa on mainly imported mineral fertilisers. That import bill and its effect on the country's balance of payments could be wiped out by using recycled waste compost as mineral fertiliser.

Back in the early 90s, the *Enterprise Initiative* of the DTI funded a series of studies looking at recycling urban wastes to land. Some of these studies ran costs into eight figures but there was one which ran a total under £40,000 before the progress generated began to become self-funding – it resulted in the farmer-owner consortium *Land Network* which, between the individual members, has, at some point, recycled nearly all the materials in the *Appendix* here; successfully, safely and within the regulations in force at the time.

The great joy of the reverse franchise is that it can be used anywhere, at any skill level of local farms and used to lead the management of food production upwards.

Part of that original study looked at how much 'waste' there might be nationally which could be recycled to land sustainably. The figures were potentially unreliable but, after many discussions, including with what was then the Centre of Waste and Pollution Research at the University of Hull, the study concluded that there was possibly 100 million tonnes per annum. Land Reseaarch now concludes that the figure is higher, potentially much higher. The total land area of the UK is just over 24 million ha but less than 20% is arable and just over 50% grassland (productive grass, of course, requiring, high nitrogen fertiliser input). Forestry would be more productive if compost were applied, too. Therefore, around ten million hectares could be used for compost substitution for mineral fertilisers. At the (misguided and unscientific) regulatory limit of twenty-five tonnes per hectare of compost, bearing in mind that composting loses maybe a quarter of its weight, that means that the available land could use in the region of, say thirty tonnes per ha of feedstock and a total of, possibly, 300 million tonnes. Bearing in mind that there is no environmental risk in applying an unlimited amount of well made compost per ha, that figure could be much higher.

Alternative Waste Treatments and Biofuels

The example above, of waste into compost, fertilise an oilseed crop and produce biodiesel on the same farm has happened. The figures can be argued about, but not the principle. The basic principle is simple enough; use wastes to fertilise a crop which involves chlorophyll (in a green leaf or algae) to convert sunlight to a usable energy source.

The waste treatment can be direct to land, it can be composting, it can be anaerobic digestion, pyrolysis or any other process which we already know about or may come up brand new. However, the bit that matters is sunlight on chlorophyll. We know this works as a mechanism. We know how to manage it. It can be used globally. It has been done before.[163]

Various estimates have been made of how much land would be necessary to produce enough biofuels for the world's cars, trucks or aeroplanes. Two thirds of the globe is covered by sea and we are a long way off managing that with green algae production and the potential management risks of using this route are very significant in terms of pollution risk simply because of tides, winds, currents and water turbulence. The safe route is land. The bad news about land is that *they have stopped making it*. The good news is that there are large tracts, very large tracts, which are not cropped. However, much of these areas are difficult to farm for many reasons and much is desert (see *Chapter 21*).

Conventional and Zero Tillage Systems

More than half the crops in the USA are planted using zero tillage, i.e. the seed drill plants seed into the ground without prior cultivation. In the UK, that way of planting is generally called *direct drilling* which was developed by ICI Plant Protection Ltd during the 1970s and 80s using the translocated green leaf plant killer called gramoxone. At that time, the percentage of the national crop planted this way may have approached the current USA figure but whatever that figure was, the figure fell back to low single figures in the early years of this century. Now, maybe half of all crops in the UK are sown with some sort of reduced cultivations but not as little as direct drilling or zero till. It is relevant to ask what was the difference in results in terms of crop yield? There is a difference in that, nationally, average yields in the UK are generally twice that of the USA. The reasons for that difference are, of course, not just a question of tillage system, it is very much more complicated than that. Nevertheless, this again raises the question of the relationship of energy inputs, financial returns and emissions of carbon dioxide and other GHGs.

Back in the 1970s, as a senior lecturer and supervisor of the post-grad *National Diploma* in agricultural engineering in England, I was involved in much of the research into cultivation inputs and farm incomes. Part of this research was involved with the burning of straw and direct drilling (zero till) compared with traditional cultivation systems. My book, *The Straw Manual*[164] covered some of this research and one example is shown in fig 20.2 opposite.

---

163. Mann, CC: *Our Good Earth*, *National Geographic*, September 2008, pp80–107

164. Butterworth, B: *The Straw Manual*, *Spon*, 1986

Energy inputs into different cultivation systems

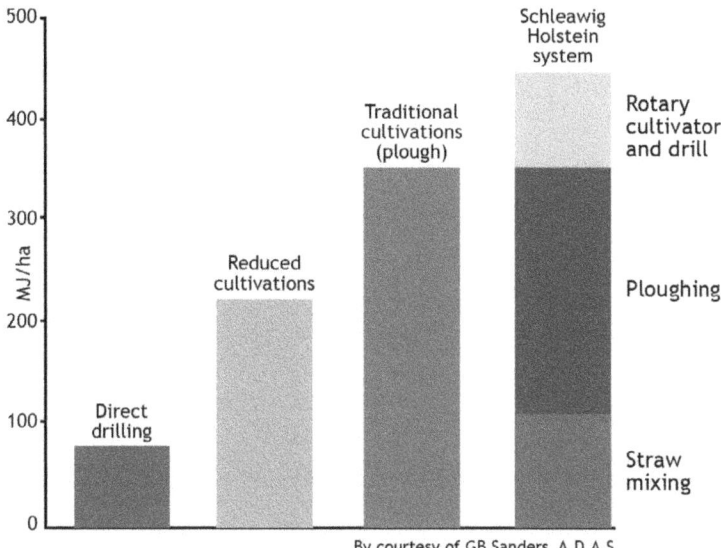

Fig 20.2: Energy Inputs into Different Cultivation Systems

So, it is possible to have different energy inputs which, in turn, have lower carbon dioxide emissions. However, it is interesting to look at this in a bit more detail. It is worth quoting some more detailed work carried out around the same time at the NIAE (National Institute of Agricultural Engineering, Silsoe in the UK). Fig 20.3 below makes a series of comparisons and it is worth putting system number eight against all the others. Direct drilling, i.e. zero till, has dramatically less inputs (both energy and labour) than any of the other systems.

Fig. 20.3 adds more data including rates of work which will, of course, affect timeliness and eventually crop yield. In the UK at that time, soil moisture retention was not of great interest but in many situations it would now be considered important or even vital to produce a crop and profit.

Again, in that same book *The Straw Manual*,[165] many different bits of research were reported on and the same conclusion came out; there was very little difference in crop yield from the various cultivation systems studied. If anything, zero till, direct drilling, came out slightly better yield and, certainly, for significantly lower inputs. There is, however, one very relevant factor in all of these studies which is not recorded directly. In each case, there was a very high level of technical input and managerial time put into each and every part of the trial and each cultivation system. It is important to note that direct drilling takes just as much husbandry time and skill as any other system, it is not a quick way out of a muddle. One of the biggest husbandry needs of direct drilling is to be much more careful about soil compaction.

---

165. Ibid.

A comparison of culivation systems, energy inputs and yields

(Grain yield in tonnes/ha corrected to 15% m.c. Straw incorporated at 4.5 tonnes of baleable straw/ha estimated at total straw yield of 9.0 tonne/ha)

	Plants/m²		Fertile		
	Autumn 1983	Spring 1984	Ear nos/m² July '84	Fertile ears/plant July '84	Yield
Plough 20cm	280	190	604	3.1	8.74
Plough 10cm	285	194	535	2.7	8.75
Tine 15cm	247	184	603	3.2	8.44
Tine 10cm	237	171	600	3.5	8.48
Rotadigger	266	174	662	3.8	9.09
Dynadrive	204	142	556	3.9	8.31
Tillaerator	221	149	630	4.4	8.52
Rotavate and plough 20cm	270	186	526	2.8	8.77
Burn and direct-drill	263	280	625	2.2	9.19
Burn and plough 20cm	298	193	510	2.6	8.84

Least significant is 0.72 tonnes per hectare, i.e. none of the yield difference are statistically significant.

Fig 20.3: Results Immediately After 1984 Harvest at Bridgets EHF for Winter Wheat

There is one other area of empirical research I directed in the 1980s which related all of this sort of academically supervised field studies to farming reality in commercial operations. This was a competition for ICI Plant Protection (as was at the time) in the study of the use of gramoxone, the translocated green leaf herbicide. This involved a competition between fifty farmers who had a wide spread of soil types and systems. They were asked to choose one field on which they cut the number of passes by one and everything else was to remain as normal. All passes to establish a crop was independently costed to a common standard and yield on the one field with reduced pass compared to the yield on other fields on the same farm. The yields were valued, again against a common standard. This allowed a *margin over establishment costs* (MOEC) to be calculated. The average MOEC improvement by a single pass reduction was 19% but the competition winner reduced by 26%. While the competition added a bit of fun, there was, of course, both commercial and serious research targets in this operation and there were several interesting points about this in-field research. Firstly, no one in the organisation of this project advised, or even hinted at what the farmer should alter; the farmer made his own choices. Importantly, they all cut different things. That is maybe the most important of all factors in all of this research work. There is no one, no one, who knows the circumstances in any particular field than the farmer himself. It is his/her skill that makes or breaks a situation. Whatever the text book says, it is whether the man actually in charge is good or not – and none of us are good all the time because nature is always ahead of us. The science and technology are vital but not much use without the experience and intuition of the manager. Secondly. When the number of passes on one field went down, the yields on *other* fields often went up. The reason was simple enough, timeliness on the other fields improved and therefore yields followed.

There are some very clear conclusions from all of this discussion. Firstly, we have known for decades that yields were not directly related to energy inputs; work the ground more does not always mean that yields follow. There is no doubt from the research described earlier in this chapter that where the management input into the field is consistent, then all the systems came out roughly the same yields. A good jockey can make a bad horse win. The farmers in the above competition all cut the number of passes on a field – but they chose which one and the top performers were thinking about what and when moving towards doing what that situation demanded on the day and making it work.

It is worth going back to fundamentals. In a natural ecosystem indigenous plants grow and thrive without cultivations; so why do farmers need to cultivate? Certainly, seed can be broadcast onto an uncultivated soils surface. If this is the situation, the percentage establishment is likely to be very low because of losses to predators, drought, poor penetration of the surface and so on. So, cultivation to plant a seed has some merit but that cultivation is very small involving only a tiny slit.

Secondly, there may be barriers to root growth and these barriers may be a result of several causes. Some soils, such as those called *podsols* have natural barriers created by acidity dissolving chemicals in the top soil and then depositing them lower down as a result of changes in acidity and thus forming a very hard, almost impervious layer. To farm these soils, regular pan busting is necessary. However, most of the barriers to root growth in established farm soils are man-made, mostly by heavy machinery. There has been quite a bit of research around the world on 'zero traffic' systems in which no machinery travels on the cropped area. A simple example of this is the beds in glasshouses and a field example is that of growing potatoes and some vegetables in wide beds. However, these systems are usually maintained for one year only. The use of tram-lined drilling gives a marker for the following crop sprayer. Some farms in the UK did try putting permanent markers at field boundaries, so sacrificing the tram-lined soil to a permanent potion but minimising compaction elsewhere. Now we have accurate satellite navigation systems which can do all these things automatically.

However, there is still the problem of compaction from combines and grain trailers. Part of the answer here is to fit the combines with tracks rather than wheels and leave grain trailers at headlands. Some farms cultivate a field, leaving the headlands uncultivated, and plant up the field. When that field planting is done, then, and only then, cultivate the headlands and plant them up. This does involve more moving about of machinery, which takes time and resources, but it does reduce compaction and that is one of the many factors directly related to yield. This discussion underlines the care needed to reduce compaction as much as possible, especially when considering direct drilling.

The third reason for cultivating is that, at least historically, it was necessary to kill weeds and this is still a valid reason to cultivate. However, the weed seed population in most farm soils is very large and cultivation scratches the surface of weed seeds and may result in more weeds, not less. Alternatively, herbicides are very good at controlling weeds and they are comparatively fast and low cost. Nevertheless, herbicides may be successful for a number of years and gradually fail against some weeds, especially grass weeds in cereals.

Case study

It is worthy of note that in the field research I carried out for ICI Plant Protection as discussed above, I visited a farm where there was a five year rotation; four years winter wheat and one year oil seed rape. Black grass became a problem which the available chemicals at the time would not control adequately. The solution was to harvest the wheat for four years, spray off the weeds before direct drilling the next crop. However, at the end of harvest of oil seed rape in the fifth year oil seed rape is an early harvest, so there was more time available) he deep ploughed and then went back into wheat. The farmer noticed that each year that at least 80% of the black grass seeds survived spraying and the direct drill and, therefore, the weed infestation got progressively worse. However, after deep ploughing, over 80% of the black grass seeds did not survive and were dead by the time of the next ploughing in five years' time. The point here is that this farmer did what was necessary using the best available technology at the time, and did not do more than that.

The overall conclusion is that in any particular field situation, it may be necessary to sub-soil, or deep plough out ruts from last year's wet harvest, or give an extra spray pass to kill a particular pest, or any other necessary adjustment, *BUT* there is little evidence to support the idea that a particular cultivation practice ensures success or is necessary. Zero till can and does work with equal yield in many situations, as so do more complicated, multi-pass cultivation systems. Generally, the more energy put in, the longer the time and the less the timeliness and this will in most situations affect crop yield. Reducing the energy inputs and moving wherever possible to full zero till is where global production has to go and makes sense for the individual farmer.

The reason for this discussion in the present context is uncomplicated; lower energy inputs mean less carbon dioxide from tractor engines and less oxidation of soil organic matter – a double benefit.

There is one more factor of fundamental significance and it is behind much of this section on reducing energy inputs into cultivations. As soil organic matter falls, usually because of cultivations,[166,167] so shear strength of the soil increases, it is a sharply inverse relationship.[168] It follows that raising the organic matter will reduce the energy input required by any given cultivation operation and may remove the need for it. High organic matter soils distort less under load and have better gas exchange and water managing characteristics. The more organic matter, the more likely it is that a move to zero tillage will work and give the highest yields.

Spreading to Land

Generally speaking, if any material is spread out far enough and enough time is given, soils will deal with anything, however 'toxic'; this is known as *dispersion technology*. In farming, they have been spreading muck for centuries. Farmers have an instinctive understanding

166. Sara F Wright & Kristine A Nichols are with the SDAARS. This research is part of *Soil Resource Management* and *ARS National Program* (#202)

167. Gordon Spoor was a lecturer and researcher at the National College of Agriculture Engineering, Silsoe, UK, for more than twenty years in the 1980s and 1990s. He was acknowledged as a world expert on soil strengths and published many papers on the subject

168. Ibid.

of what is good for their land. It is in the nature of regulators that they are unlikely to understand this and, therefore, will not easily trust. However, regulators do have a point. While the farmer may be trusted up to a point, and that local, in-depth experience is never anything less than invaluable, it is also necessary to recognise that wastes may not be what they used to be and that the regulator has a responsibility, too. Despite that recognition, there is a difficulty in that most regulators are not hands-on agricultural scientists. It is inevitably true that responsibility without adequate technological training for the job will result in inhibition and a slowdown in activity and serious difficulty for the innovative or entrepreneurial. This is one of the fundamental weaknesses of western democracy. The answer is to license operators on the basis of academic and technological training and allow easy routes into stepped operation with a withdrawal of the licence if and when pollution occurs.

Liquid wastes

It is important to remember that the compost process needs the micro-organisms, the food, oxygen and *moisture*. Green waste on its own, over a year's supply of materials including relatively woody inputs, will certainly go black and friable in most circumstances, but the process is more likely than not to be incomplete with much of the potentially soluble nutrients still soluble. It is only by thorough composting, which needs moisture to push it all the way through its process, that solubles will turn into humus and reduce the leachable nutrients to near zero; *that* is the way to protect groundwater after spreading.

One of the reasons for adopting *deep clamp* composting is that it is comparatively safe to add liquids to the top of a 3m deep heap with very low risk of leachate from the added liquid running through and out of the base of the heap. Regulators with little practical experience of composting will generally argue that the deep clamp still needs to be placed on concrete in order to eliminate the point risk. Concrete has significant energy cost and it really is not necessary when a competent operator is given a reasonable code of practice for additions of liquids to deep clamp operations. Even if those conditions of competence are not where they could and should be, the risks on small scale operations are still very low and are generally out-weighed by the advantages.

If, for whatever reason, the addition of liquids to compost is not an option, then many materials (part composted or 'fresh') can, safely and at low cost, be spread direct to land. To get an even spread and avoid odour, the likely best machine will be a field tanker equipped with dribble bar (the *dribble bar* is a wide boom with trailing, large-bore flexible pipes trailing onto or just above the ground). The bad news about this method is the labour cost and the weight on the ground which, in wet conditions on many soils, will rule it out. An alternative is to use a reel-type irrigator and nurse tank. Again, it is possible to equip the light-weight field bogey with rain gun (which will maximise output and odour control), a boom with sprinklers, or again, a dribble bar (which gives the low odour advantage).

Biosolids: Sludge and Cake

- *Sewage sludge* is a phrase with many meanings. In practice, very little original-state (or 'raw') sewage gets direct to land in Europe, and many other developed-economy countries, too. When sewage material undergoes a significant level of treatment, it is better and more correctly named *biosolids*.

- Biosolids sludge is a pumpable liquid with dry matter content usually around two or three or maybe even over ten per cent (highly variable in practice). As the dry matter content rises towards 20%, the material becomes increasingly stackable. From around 18% it may reasonably be called *cake*.

Nutrient Values
Generally speaking, biosolids are useful in terms of nitrogen and phosphates but short of potash. Any particular sample can be analysed to give a specific guide. Generally and within the regulations, where biosolids can be applied, they can be applied at a rate which will supply all the nitrogen and phosphate needs of most crops. Because of the high level biological activity encouraged by the material, it may be that that activity will facilitate the release of otherwise immobile, or 'unavailable' potash. The risks are discounted in an obvious and logical way. For example, raw, untreated sewage is not allowed. Biosolids which have been subject to secondary treatment (such as mesophilic digestion which takes the material up to 38°C) can only be spread to food crops which might be eaten raw (such as salads) up to thirty months before harvest, applied to vegetables which will be cooked up to twelve months before harvest, and to industrial crops (such as oil seed rape for biofuel production) anytime. Further, if the biosolids have been subject to tertiary level treatment (such as thermophilic digestion which takes the material up to 60°C), well, that can go anywhere, anytime.

- The system works because it is seen to handle and provide for risk in a clear and workable way.

- There is no reason why this approach cannot be used for the output from MBT (mechanical and biological treatment) plants. For many years, up to early 2009, Defra prevaricated and avoided allowing the output of an MBT plant to go to farm land, unless the input to the MBT plant was source-separated (which is the whole idea of an MBT that it can take anything and everything, so avoiding source-separation which is expensive, untrustworthy and the public do not like it). It can go to cap out a landfill (so, presumably, there is no risk to the environment??). Now, Defra has the compost standard of PAS100. It is either good enough or it is not. So either adjust it or use it for MBT output. This lack of vision and decisive action to help recycling forward is, at best, sad.

Protecting Groundwater & Aquifers

All organic matter has an ability to hold water. Peat will hold up to sixteen times its own weight. Most composts somewhat less – from five to ten times. Biosolids materials are able to push composts and soils well towards the moisture-holding capacity of peat. That has advantages in crop production in reducing drought stress and irrigation need. There are also potentially big advantages in run-off and flood control. These *top soil reservoirs* can be built and maintained by adding organic matter in large quantities. Large quantities can be nitrate-safe. One such source of top soil reservoir building organic matter is already within the hands of the water companies; one million dry tonnes of output from the STWs (sewage treatment works) in the UK. This will probably give a reservoir effect in an agricultural soil of in excess of five million tonnes of water.

That original dry tonnage starts off as twenty-five million tonnes of brown water. If it were thermophilically digested, it would produce methane gas which is a useful biofuel. The residue liquor could be used on arable farms nearby where it could be piped to farmland close by the point of production. De-fibred sludge could be spread direct by rain guns or dribble bar or stored for a suitable crop window. There are clear advantages to the logistics interests of all parties.

A press report (by the author of this book) in the late 1990s quoted the following:

> Some years agao, I met Charles Booth farming 174 ha of arable land near Knottingley, Yorkshire, UK. He has used 4% sludge for twenty years. He "would not like to farm without it". He readily claims lower fertiliser costs, less crop disease and higher yields because of a planned use of biosolids. It is not that he 'thinks' that; he has direct, academically credible evidence because Ryhill Farm Services, a crop protection chemical distributor, ran an agronomy centre on the farm with properly run trials. "As an example, one twenty ha field had 3920 m³ of sludge applied last August, just before ploughing. That added the equivalent of over half an inch of rain. The land was easier to work and the crop of winter wheat established quickly with less risk. Because of a healthy, vigorous crop, I expect at least two tonnes more yield per ha than my neighbours." Independently assessed figures show that this farm has 20% more yield than its neighbours and 80% less fertiliser cost.

The fact is that, partly unknowingly, the water recycling industry is already involved in managing top soil reservoirs. It is also involved in very significant environmental safety advantages compared with other methods of acquiring and applying fertiliser. Biosolids products release plant nutrients slowly, when the crop can take them up, so leaching and pollution of underground water is less than with mineral fertilisers. There is growing evidence of both anti-biotic and pro-biotic effects, so producing healthier crops with less disease and less use of spray chemicals. There is good evidence of better, more consistent yields for the farmer partly because of the soil reservoir effect reducing drought stress on the crop.

Reservoirs & Dams

David Setler, a researcher in Sri Lanka, said, "Everywhere in the world, there is a need for more reservoirs and dams." Research in the USA gives good, practical data on do-it-yourself (on the farm) reservoir building using old motor tyres as fill for dams and retaining banks.[169]

There is a compelling, common sense argument to build capacity where it is needed. Neither on-farm conventional, nor soil reservoirs have the potential to solve all the demand problems but there is scope to make a real and dramatic contribution by reduction of run-off and evaporation. Much of the technology, experience and culture is already inside the water recycling industry. Holding water in the uplands can reduce, even prevent, flash flooding lower down, and general flooding in the lowlands. The problem is recognising that fact and using it to advantage. For the first time in fifty years, farming is really ready to help – very actively, if managed right.

---

169. Hoenig, SA: *The Use of Used Tyres in Water Systems, USDA, ARS 173*

Deforestation, Overgrazing & Soil Destruction

Much of the USA is still covered in virgin forest. Conversely while much of the Amazon basin of South America is still virgin rainforest, much has been cleared for cropping. 5000 years ago, the Sahara Desert was covered in trees. Similarly nearly all of Britain and Ireland was covered in trees. In the USA, there have been dust bowls where there had been productive farming and much of upland Britain is now bare rock.[170]

When humans first settle on the land, they grow livestock and cut trees down. Then man graduated to cultivating the ground and growing crops. They next cut more trees down and run down soil organic matter. Soils then erode by wind and water. In the UK, something of the order of two to three million tonnes of top soil get lost in the rivers to sea every year. Estimates of erosion globally vary but it will be several million tonnes. It is quite simple, deforestation and overgrazing destroys soil. As Roosevelt said, "a nation which destroys its soil destroys itself."

Climate Change
Because farming uses such a significant proportion of the world's use of energy, farming also similarly makes a substantial contribution to global warming. Farming can cut that use, most obviously by reducing cultivations and use of mineral fertilisers. The farming industry can also do two very positive things. Firstly, by its very nature, the green leaf takes carbon dioxide out of the atmosphere and gives us oxygen in return. As an industry, farming should be much more vocal about that. Secondly, recycling waste to farm land not only puts the carbon in that waste into the soil organic carbon bank, it saves the use of mineral fertilisers.

Flood Control

A code of practice agreed between the Environment Agency and a co-operative of farmers in an upland area could dramatically affect lowland flooding.

Few in the UK can be unaware of the flooding in many parts of the country over the last few winters. What always follows are discussions of why and who to blame, mixed with theories which are often held passionately as the silver bullet to cure it in future. The truth is that the solutions are complex and a belief in a single solution indicates a lack of understanding of the natural environment.

No doubt, the dredging of water escape routes, of the building of 'sacrifice' areas, building of new extra drains, and all the other construction possibilities are part of the defence for future urban protection. The holding of water in the uplands in order to give slower release is the subject of what follows here.

Firstly, the farming of the area is the key to the management of the uplands, not the cause of lowland problems. More specifically, if any change is not economic for the farmers involved, then either the taxpayers come up with the cash, or it will not happen at all. So, finding a development which slows run-off and is financially attractive for farming to produce food is the only acceptable way forward that has a chance of working. There is a way to do this and it has already been done.

We already know that high organic matter soils hold water better. To put figures on it; sand will hold its own weight of water, clay twice its own weight but composts will

170. Millward, R & Robinson, A: *Upland Britain*, *David and Charles London*, 1980

hold five to sixteen times their own weight. We also know that bare soil (without an established crop on it) erodes easily. So, enabling farmers to develop high organic matter soils and grow crops with a minimum of bare soil will improve matters in the lowlands. We already know much about composting urban wastes and about forestry and other crops which can reduce erosion. Composting wastes can be very profitable but there is always an assumption that compost from wastes will be spread on food-producing land. There is always a lot of paperwork, some of which may be a bit counter-productive. Suppose a group of farmers were to get together with the Environment Agency and develop an area plan to maximise the volume and type of 'wastes' which could be composted and spread and, instead of lots of individual applications to spread, a code of practice for food and forestry land? Planting tree belts would help, too.

Reforestation
...and Global Warming Black Death, Chinese Discovery of the World

From an environmental point of view, it does not matter what the crop might be, either the compost is fit to put on a particular soil, or it is not fit. However, public images and regulators being what they are, if a particular compost or 'waste' is being put on food-producing land, then the attitude will not be the same if that material were to be put on forestry land. So, forestry may allow a wider range of materials or processes to be used. Furthermore, trees are good at taking carbon dioxide out of the atmosphere and timber is a crop which will grow ion places where normal farm cropping such as wheat, will not grow.

Could we really grow trees on the tops of the UK's highest mountains and make the Sahara Desert bloom? Yes we can. We know this because 5000 years ago the whole of the UK used to be covered in trees but they were cut down to build ships, the land was grazed for wool production and over-grazed; the result was the washing of soils into the sea. That land could be reclaimed with soil based on wastes and re-forested. That could be done on land up to 2500 ft (770 metres) of UK mountains and on the sands of the Sahara.

Right back at the beginning of this book (see *Section 2*) there was reference to the Black Death in Europe and its effect on forestation and global temperatures. The University of Utrecht[171] found reason to link the dramatic reduction of population following a series of pandemics of 'the plague', the abandonment of agricultural land, natural reforestation of that land, and changes in global temperatures by around one degree Centigrade. It is worth looking back at the actual dates and population changes, put by many but none more dramatically than William Stanton. Population rose during the Roman occupation of the British Isles up until between 300 and 400 AD. There was a fall, due to conflict, disease and slavery, in population between 400 and the arrival of the Normans in 1066 after which the stability allowed population to rise again. That rise continued until the Black Death came out of Asia and, between 1348 and 1400 cut the population of Britain from around five million to around two million. Now, just hold onto those dates and remember the Utrecht research linking falls in population to re-afforestation and reduction of global temperatures.

171. van Hoof, TB et al: *Forest Re-Growth on Medieval Farmland After the Black Death Pandemic – Implications for Atmospheric CO2 Levels, Elsevier*, February 2nd 2006

Refeering back to the work undertaken by Gavin Menzies in *1421 – The year China Discovered the World*, the author looked in detail at the vision of Emperor Tsu Di who sent his admirals to map the world and bring it all into China's tribute system of trading. As Menzies detailed, the Chinese 'treasure' fleets did circumnavigate the world and discovered the Americas and Australia long before the Europeans. These fleets were truly enormous with the largest ships over 400 feet long (125 metres) and, the larger fleets over 800 vessels. These were ocean-capable craft and made of teak. Menzies lists five admirals, each with a fleet. As a matter of arithmetic, Emperor Tsu Di must have ordered the construction of 3000 to 4000 ships of various sizes, but not small as they were to be ocean-going. The Chinese had also had a merchant fleet for trading in the China Seas and Indian Ocean for several hundred years. In the ten to fifteen years before the admirals set off on their voyages, there must have been construction of maybe 5000 ships. He also built the Imperial Palace and many other buildings from teak... *all* from teak. That must have involved the destruction of an enormous number of trees. Also, some 5000 years ago, the Sahara Desert was covered in trees.

There is no doubt that the loss of tree cover on a global scale is related to global warming and its control.

It is not the purpose of this book to look more at the tree issue other than to observe that the earth used to have rather more of them and they are rather good at taking carbon dioxide out of the atmosphere *and* pumping oxygen back in. So, reclaiming the deserts and planting trees and/or growing food may be an issue worth looking at.

Food Production & Human Health

The most difficult question which Land Research has to wrestle with is not how to monitor these elements in the soil and in the compost of direct spread waste, it is how to do this economically with the resources which a small commercially-driven organisation can risk committing. The technology and experience does exist but the trained manpower is limited and significant resources are needed. Nevertheless, progress is being made. Safety is not just about 'dilution', nor is it just about spreading it out so that nature can cope. It is partly about what has **not** been put on the land. Some arable soils have had no organic manures, no animal waste, no compost, just relatively pure mineral fertilisers, added for over fifty years. Ammonium nitrate is just ammonium nitrate – no trace elements. However, in that time, harvested crops will have removed enormous amounts of trace elements.

This, then, defines the big challenge for recycling to land. We have to replace these trace elements. On evidence, it seems unavoidable that this is a matter of human health and longevity. The most economical way to do this is via 'waste', which also appeals as being 'natural'. We also have to replace organic matter (large carbon-based molecules). That means that we can use plastics – not robust plastic sheets (such as polyethylene or *polythene*) but liquids which can be incorporated and spread out onto every particle in the compost mass. We also need fibres in the soil to manage moisture and gas exchange. There will be less flash flooding if we have fibres to hold the surface open. If the land is subject to intensive cultivation, these fibres may deteriorate (by oxidation) too rapidly and synthetic fibres such as carpets (which have been finely shredded) can and will provide that physical function.

Reclamation of Land to Grow Energy Crops

There is, of course, an area of land which does not have the complication of existing crop; that of reclamation. Reclamation may be from previous industrial activity or from desert or soil-depleted or soil-less upland (although that, too, is quite likely to be from previous human activity.

That, from a recycling waste point of view, is clearly an attractive option. However, as chapters in this book have shown, there is also a compelling attraction of this in terms of environmental stability and the very survival of the human race. We really do need to look at the reforestation of this land, preferably for trees and crops which are energy-producing.

Manufacturing Soils

Fertile soils are made up of a mineral fraction, a chemical fraction including both inorganic and organic materials and a biological fraction. While the formation of soils in the natural state took many millions of years, it is possible to manufacture soils somewhat faster – but not instantly.

It is possible to make a soil from 'wastes' which will, over time, be at least satisfactory, and possibly really good. There are three basic fundamentals: the physical components and structure, the chemical constituents and the biological activity.

The physical structure of soils involves mineral particles (sand, silt and clay) in any proportion and containing stones and all sorts of 'foreign' bodies. This structure needs to present an anchorage into which roots can (reasonably easily) grow. In uncultivated soils, this may, in time, become rather like a stack of uneven concrete blocks with the gas exchange, moisture movement and root penetration mainly in the place where the mortar would be in a wall. If these movements are restricted, then it does not work so well. In a heavily cultivated soil, this structure will break down and future cultivations will need more power, the crop will need more nutrients and each plant may become more susceptible to disease. High-tech farming can go a long way to living with these problems.

A chemical structure which provides a balance of crop nutrients is the second essential. The absence of enough of the right nutrients in the right proportions to each other will result in a crop limited in its growth and more susceptible to stress from disease, drought, water-logging and wildlife attack. The range and balance of what these nutrients might be can be studied in many other texts. What this text has concentrated on is the contribution to crop nutrition by composts and refers to the work done by Albrecht and Kinsey.[172,173] It is not just the nutrients as a list, nor just the individual supply of each nutrient, but also the balance of these elements and how this is related to the physical structure of the soil and the presence of humus.

---

172. Kinsey, N: *Hands on Agronomy*, <u>Acres USA</u>, 1999

173. Mann, CC: *Our Good Earth*, <u>National Geographic</u>, September 2008, pp80–107

The biological activity in soils is equally important as the physical and chemical status of the soil. The evidence is that crops are fed substantially through the mycorrhiza,[174,175,176] certainly so in natural ecosystems,[177,178] or compost-based crop production. In mineral fertiliser-based crop cultivation, it is likely that soluble materials usually enter the crop roots directly and in solution, avoiding the mycorrhiza at least at least in part and probably mainly. The evidence in this text points to greater stability to the farmer and a move on all fronts towards true sustainability, by moving to humus/mycorrhiza-based soil management. This means that the mycorrhiza need feeding and, fortunately, they can be managed much as crops are managed with mineral fertilisers, except that the nutrients need to come via humus and carbon chain molecules.[179]

So, in temperate climates, we can take construction waste 'fines' (the sandy materials passing through a screen) with the right chemical composition, add a suitable compost (again with the right chemical composition) and produce a potentially productive soil. Suppose we apply this thought to reclaiming erosion-depleted upland soils or to making the deserts of the arid areas of the world productive – as many used to be? As an example, Malta is a relatively small island with a maximum distance coast to coast of only seven times its airport runway length. Malta has around 400,000 total resident population and around two million tourists a year. Despite this population pressure, over 70% of the nation's waste output is construction waste. The island was almost certainly covered in much more vegetation historically and yet much of the surface of the island, that which is not covered in concrete, is eroded back to the bare rock or, at best, very thin soils except in a limited amount of agricultural soils (which are still quite shallow and mostly poor in organic matter). That bare rock could be recovered back to vegetation and trees by using crushed and screened (or just screened) construction waste plus composted organic wastes and biosolids (the large particles from screening could be used for wind-protection bunds, inland dams or sea walls).

The same approach could be used in the UK to reclaim uplands denuded of soil by over-grazing in the past; similarly in most of the countries of the developed world.

Urban Soils

The University of Doncaster in the UK published a 'policy briefing' document in 2015 referring to the lack of interest in urban soils and the place they take in preserving the wider stable environment. It is worth quoting:

174. Butterworth, B: *How to Make On-Farm Composting Work*, *MX Publishing*, London, 2009

175. Butterworth, B: *Managing the Soil Rumen and Ion Exchange*, *Arable Farming*, September 9 2000

176. Butterworth, B: *A Top Idea That Holds Water*, *Water and Effluent Treatment News*, October 6th 1999

177. Sara F Wright & Kristine A Nichols are with the SDAARS. This research is part of *Soil Resource Management* and *ARS National Program* (#202)

178. McKenzie, D: *Facing the Resistance*, *New Scientist*, January 19th 2019, pp20–21

179. Butterworth, B: *How to Make On-Farm Composting Work*, *MX Publishing*, London, 2009

Urban soils are an essential non-renewable resource and an often overlooked determinant of environmental and public health. They provide a vital support mechanism for the constructed environment as well as the enormous variety of green infrastructures in cities. They have the potential to help mitigate flooding, improve community wellbeing, increase biodiversity and store carbon, all of which ultimately help in the fight against climate change.

The water holding capacity of a soil and therefore its flood and drought resistance relies on its organic carbon content. These carbon levels are rapidly reducing due to climate change and as organic wastes are produced during food and textile manufacturing processes but not returned to the soil. A 2014 report published by the House of Lords Science and Technology Select Committee identifies thirty million tonnes of organic waste produced each year in the UK that have come from the land but are currently not returned to it.

This makes soil more susceptible to the damaging effects of climate change in future, resulting in a vicious cycle of soil degradation. The health of our soils in urban areas is central to the mitigation of flood risk. While floods can affect all communities, the vast majority of UK population lives in an urban area. This means that urban soils are particularly vulnerable to degradation and are consequently disproportionately affected by increased rainfall as a result of climate change.

The Consequences of Neglecting Our Urban Soils

The total annual cost of soil degradation in England and Wales is likely to sit at £1.2 billion a year. The social and economic devastation caused by flooding is readily apparent. But the issue hidden in plain sight is that the muddy flood waters are taking even more soil organic matter and minerals out of our soils, exacerbating the vicious cycle of climate change – soil degradation – climate change – soil degradation. The devastating effects of recent flooding in our cities could have be reduced if our soils were capable of storing and transmitting water at the same time as retaining their strength so that they are not washes away.

Recent scientific studies on mineral-organic interactions conducted by Durham University highlights that minerals actively stabilise organic matter in terrestrial environments. Engineers have the skills to pre-process recycled minerals together with organic wastes to improve soil, which could improve flood resilience, store carbon and improve soil health and plant growth as well as moving us closer to a circular economy.

PCCS – Photosynthetic Carbon Capture & Storage

The Rio, Kyoto, Bali, Copenhagen, Paris and now Poland conferences, and every one since, have raised many questions. This book looks at some of the answers. There is a link between current fossilised fuel consumption, desertification, urban waste and biofuels. The real question, then, is how can this link be identified and turned into a business with real jobs, real income and, ultimately for governments, tax revenues.

As the pressure in middle-class Western society has been formalised at the various 'environmental summits', increases its opposition to the chopping down of rain forests to make biofuels, the converse becomes more of an opportunity; to buy biofuels made from crops grown in arid climates and on desert soils. By the late twentieth century, i.e. the late 1900s, energy and how long the fossilised fuel reserves would last, had become significant issues. By around 2010, the general consensus had fundamentally changed; everyone thought that there was enough fuel reserves but that pollution and rising levels of carbon dioxide in the atmosphere were now compelling issues. The idea that all oil and gas resources have a finite volume and that, progressively, they will become more valuable until they actually run out, was less important but pollution was the real issue. There is, then, a compelling logic in investing a proportion of oil revenues in reclaiming desert for biofuel production. The one big advantage of biofuels produced from cropping on the land is that the green leaf, by field crops or algae, takes carbon dioxide out of the atmosphere and put oxygen back in.

Recycling to Land as a Business

What, if anything, will drive this is enabling regulation and cash. Where the gate fee for alternatives is greater (which can be exaggerated by tax as in the landfill tax) then there is a motive for waste producers to go through the farm gate and pay the rate per tonne.

In the planning of an on-farm composting facility, the unknowns can, to some extent, be examined and planned for with sensitivity analysis. Sensitivity analysis (in this case relating a range of gate fees to a range of tonnages) can be related to capital and running costs. But in addition to this, it is the vision of waste integrated to its value on the land which has driven the growth and success of the Land Research farms. That is developing into sustainability not only in the narrow farm/environmental sense but into the global sense including energy sustainability.

Chapter 21: Top Soil Reservoirs

Reservoirs at the Crop Roots

Trying to grow most crops in areas where the annual rainfall is less than 380 mm has its problems; the biggest is that there is not enough water. This chapter of this book is a new look at old technology in pointing to more successful ways of developing the reservoir at the plant roots. Most arable farmers are conscious of the need to raise soil organic matter but ask where to get it. The use of the waste from each crop, sewage and municipal and industrial wastes can solve both problems of disposal and of soil water shortage. It is also commonly the case that accepting other people's waste, where the regulator allows it, may bring in very significant revenue.

New developments in dry-land technology suggest that raising the organic matter of sandy soils is likely to be worth fifty to 100 mm, maybe 150 mm, of rainfall per annum – because it keeps the water at the plant roots rather than draining or evaporating away. There are other nutrient and disease advantages. These figures have come from farm studies, almost unnoticed by formal research, and from all over the world on the disposal of waste (sewage and municipal waste) to agricultural land.

Growing crops successfully in any dry area or year does, of course, depend on many factors. For centuries, there has been much thought and effort put into irrigation. That, however, is expensive in resources, will always have at least some salination effect and there may not be enough water anyway. Extraction licenses, generally necessary in Europe and many other countries, to take water from rivers and boreholes, are often not easy to obtain from regulating authorities. There is, and will always be, increasing competition between urban needs for water and farming. There has, however, been comparatively little attention to the reservoir at the plant roots.

The most obvious limit to the soil reservoir is the depth to which the plant roots penetrate the soil. Many cereal and vegetable crops will put their main roots down near two metres in one growing season, if there is not a compaction barrier to that growth. . There is some evidence that cereals may put fine rots down to nine metres. So deep cracking of the soil to break pans and compaction is commonly practised – but not always when it should. There is a less obvious factor which is commonly neglected – the soil organic matter. The water retaining capacity of the soil depends on a number of factors including the mineral nature of the soil itself and the soil voids which are affected by cultivation and compaction. However, soil organic matter has a major effect and it can be raised by thoughtful management, or dramatically reduced by cultivation. As a guide, conventional cultivations in Europe and based on the mouldboard plough and harrowing, will oxidise up to 35% of humus per annum. Zero till, or direct drilling, will cut that to 10%.

Low organic matter soils are much less tolerant of heavy machinery and compact more easily. Further, it is the organic matter which is the main 'reservoir' for water. That water is held not only physically, as in a sponge, but some is held colloidally which means that it can be fed direct into the plant by the soil mycorrhiza (see *Chapter 12*).

There is a problem in cultivation; aeration will dramatically increase oxidation of organic matter. This rate of oxidation is faster and more destructive in hot soils and climates. So, in cultivation, a balance has to be struck; the higher the temperature and the more thorough the cultivation, the more rapid the destruction of organic matter.

In one study in the east of England, several farms involved in a wider scientific study were noticed to have abnormally high yields of sugar beet and other crops. Three of the farms, which were on Suffolk sands, yielded consistently higher than neighbours, all of which had apparently similar levels of husbandry. The high yielding farms had been using sewage sludge for between five and thirty years as their main source of fertiliser and achieved yields for sugar beet typically of sixteen tonnes to the hectare compared with neighbours at eleven tonnes. That was a 45% increase in yield accompanied by a 75% reduction in expenditure on mineral fertiliser.

There is increasing evidence of extra disease resistance of crops grown on soils fertilised with compost made from organic matter. Some work done by EcoSci Ltd of Devon in the UK, has shown that crops grown in the laboratory on green waste compost showed remarkable resistance to disease. There was 100% control of brown rot in potatoes, 100% control of club root in brassicas and typically 65% control of several cereal diseases.

Alwyn Moss, farming near Mildenhall in Suffolk UK in the late 1990s/early 2000s, has composted tens of thousands of tonnes of shredded newsprint (used as bedding for horses at the Newmarket racing stables) and applied 250 tonnes to the hectare – that takes some skill to plough in. The compost absorbed five to ten times its own weight of water. Ploughing in before the winter rain will provide a reservoir to produce a crop without irrigation in the following year. That means that this dressing of compost per ha will absorb and hold up to 2500 tonnes of water per hectare, equivalent to 250 mm of irrigation.

There is another potential advantage of some economic and environmental significance. Less run-off of mineral nitrogen into water courses will make a useful contribution to reducing water pollution.

Even in desert areas, there are potentially four sources of organic matter which may be within striking distance and relatively un-tapped; the waste from the last crop, sewage, industrial wastes and separated or whole MSW (municipal solid waste or 'bin rubbish').[180,181] In some areas, *unseparated* MSW may be the (perhaps surprising) opportunity.

Raw sewage has many problems if it were to be used on crops grown for human consumption. In most countries of the world, however, filtered and processed sewage is commonly so used. Most sewage digestion in the UK is mesophilic (carried out at up to about 38°C). New technology in thermophilic digestion (at 58°C) will provide major advantages. Firstly, it pasteurises the waste which will kill most pathogens. Secondly, it will

180. Berg, B et al: *Plant Litter, Decomposition, Humus Formation, Carbon Sequestration, Springer*, 2003. This is one example of many hundreds of references to organic matter degradation and formation of humus in the academic literature and on the web

181. Etriki, J: *Municipal Solid Waste Management and Institutions in Tripoli, Libya: Applying the Environmentally Sound Technologies (ESTs) Concept, University of Hull*, May 2013

allow co-digestion of added solids including crop waste or whole MSW. Thermophilic co-digestion yields much more gas (which can be used to drive engines and/or generate electricity). The liquor has three to ten per cent dry matter and is an excellent and safe fertiliser and, of course, it is mainly water. In the best possible situations, sewage from a small town or village can be piped to an on-farm digester, eliminating all trucking.

Inevitably, not all situations fit these ideal solutions and there are various levels of sophistication and engineering built into other approaches to utilisation on the land.

One of the counter-balances to rising population and industrialisation is the increase in waste. As wealth rises, so does waste. Perversely, as the waste goes up, so does the opportunity to use that waste, via composting, to produce crops without mineral fertilisers and to reclaim land for crop production with the green leaf taking carbon dioxide out of the atmosphere and pumping oxygen back in.

Reducing and Eliminating Irrigation

Water will become increasingly important in world politics and power; at least as important as hydrocarbon fuels but in a different way. Water is necessary to produce food. Estimates of the proportion of global water use by agriculture vary but, even after significant industrialisation in urban areas is likely to be above half, maybe two thirds, while total water use is going up and available water is (mainly due to pollution) going down. The use of mineral fertilisers, especially nitrogen fertilisers, is causing pollution of ground water. We have a problem and (to miss-quote a famous quotation) *Houston, we have an answer!*

There are ways of eliminating that pollution of ground water by mineral fertilisers and making irrigation water go ten times further – at least. The important information here concerns colloidal capacity. Sand has very little and will hold very little water and very little crop nutrients. . Clay has about twice that of sand and will hold twice as much water and some nutrients. Clay will hold cations (ammonium and metallic) fairly well and very little anions (nitrate, sulphate). Humus has maybe three or four times, maybe more, the colloidal capacity of sand. It will hold onto cations *and* anions. Further, it will hold five to ten times its own weight of water. Therefore, this material, put in the right amount and in the right place, could reduce irrigation need by a factor of five to ten times. Putting this a different way; it could be used to grow a particular crop with a tenth of the irrigation need.

So organic matter, forming humus in the soil, changes the way water and loss of nutrients can be managed (see *Chapter 12*).

So, to avoid pollution of groundwater, switch from soluble mineral fertilisers to organic-based technology. It is possible to stop the pollution from mineral fertiliser by adding it, half way through the process, to a compost operation based on organic wastes. It is possible to make irrigation water go much, much further by building what are called *top soil reservoirs* out of compost. Large amounts are needed and there is a great deal of suitable wastes produced in all human communities.

The problem with water in hot situations is that it evaporates easily and often with no benefit to crop growth. The worst evaporation loss comes from using sprinkler irrigation which maximises evaporation before the water hits the ground, let alone be used by the crop. While it may be important to remember that tis way of application may have effects on leaf temperatures and surface layer temperatures, it is still this maximising of evaporation loss which prompts an alternative thinking.

If it were possible to make fewer applications, maybe only one, of enough water to grow the crop and hold it near the crop roots, then that evaporation could be dramatically reduced. The idea of *top soil reservoirs* (TSRs)[182,183,184] is to accept that water loss by translocation is necessary to crop growth but to limit the loss of water that has not been through the plant.

TOP SOIL RESERVOIRS

SUBMERGED BED RESERVOIRS

Options
1. Plant seeds and irrigate
2. Transplant seedlings through layer

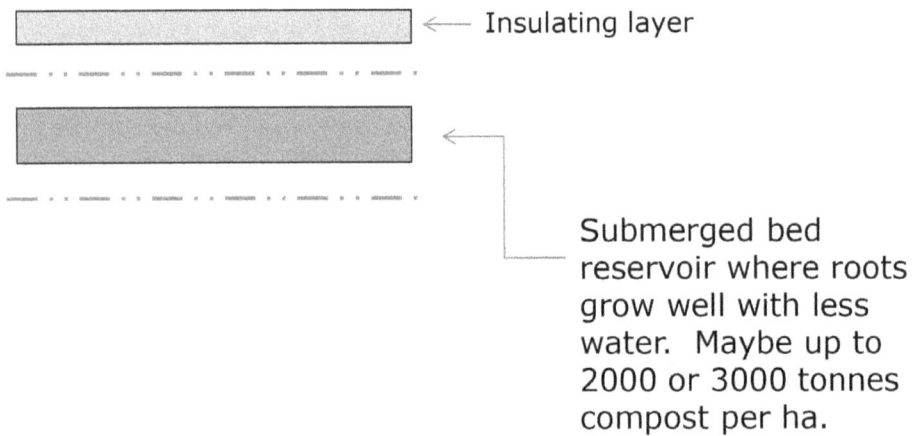

←— Insulating layer

Submerged bed reservoir where roots grow well with less water. Maybe up to 2000 or 3000 tonnes compost per ha.

Fig 21.1: The Two Layers as a Basic Concept

182. Butterworth, B: *Sewage Solution for Low-Water Farming, Far Eastern Agriculture,* September/October 1997

183. Marshall, M: *Finding the Real El Dorado, New Scientist,* Jan 19th 2019, pp26–29

184. Butterworth, B: *Reservoirs at the Crop Roots, The Vegetable Farmer,* March 1998, pp31–32

TOP SOIL RESERVOIRS

SANDY SOILS IN HIGH TEMPERATURE ENVIRONMENTS

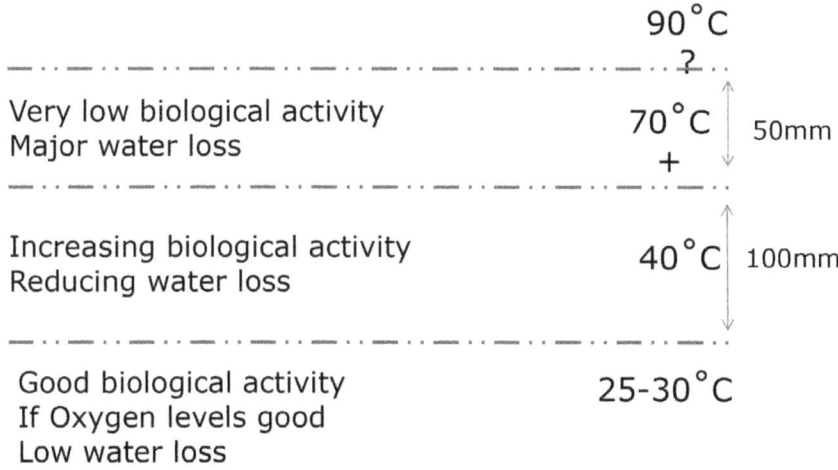

Fig 21.2: Possible and Likely Soil Temperatures in Sandy Soils in a Hot Climate

This concept indicates that the very top layer of soil is not much use in desert soils in the sense that it will often be too high a temperature for root and biological activity, lose water quickly and have its organic matter rapidly oxidised. However, it does have a real value as an insulating layer. So, the next step is to design the depth and material of that insulating layer. Fig 21.3 gives a guide.

TOP SOIL RESERVOIRS

MANAGING SOIL TEMPERATURES AND WATER CAPACITIES

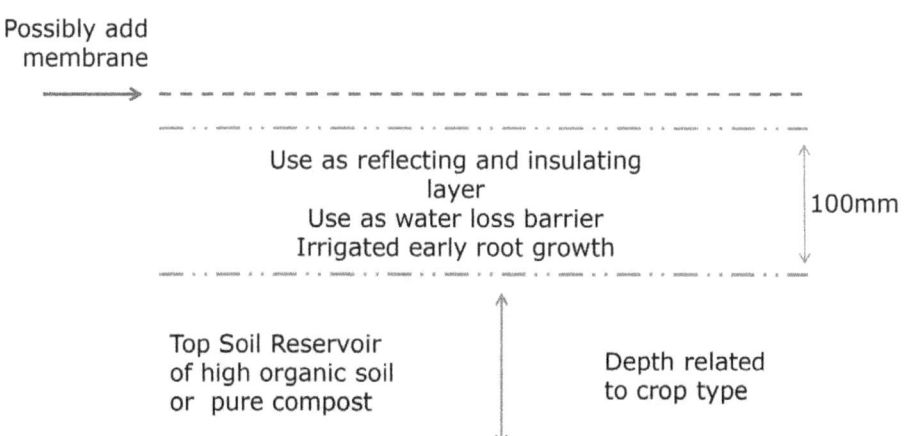

Fig 21.3: The Use of the Surface Layer as a Reflector

So, in any particular situation, there is a task which is to design to plan for a stable top layer and design and a crop establishment programme which will get crop roots down into the reservoir area. In practice, what comes first is the design of the top soil reservoir (TSR) zone itself, i.e. the zone in which productive root growth will occur. Fig 21.4 below shows the basic thinking on water holding capacities.

TOP SOIL RESERVOIRS

SOIL RESERVOIR CAPACITIES

Sand holds its own
weight of water

Top layer has almost
nil water holding

If clay: hold twice own weight of water

Composts from urban wastes will
hold 5-10 times own weight of water
AND
HOLD UNLIMITED AMOUNTS OF
PLANT NUTRIENTS
WITHOUT LOSS OR POLLUTION

Fig 21.4: Water Holding Capacities

Figure 21.5 opposite shows the principle of TSRs for any soil in any environment, temperate of desert. However, this leaves the question of how much organic matter to add per ha.

Sand is useful in all soil structures to allow movement of air and water, some clay will help colloidal capacities and, therefore, the mixture of the two plus some silt (half way between the two in particle size), i.e. 'loam' being advantageous. What is available locally will vary. Now look at the depth and construction of the TSR – the reservoir itself where the productive root growth will occur.

Many crop roots will go down to two m and the design may not only allow for that but deliberately construct that depth. That is certainly expensive and would demand very large amounts of compost. Generally, the mix of the zone will vary but, as a starting point for the sake of this discussion, say 50% local material and 50% compost. That will vary in practice depending on the calculation of what the crop needs to grow to harvest.

SOIL SURFACE CONDITIONS

Sand holds its own weight of water. If clay: hold twice own weight of water	Unstable conditions in surface layer
Composts from urban wastes will hold 5-15 times own weight of water. Store unlimited amounts of organic Carbon or bio-char here. Mix Carbon source with enough local soil to provide trafficability. (Maybe 50/50/)	Bio-Active layer Relatively stable environment. Good gas exchange. Good moisture conditions. Good biological activity. Good root growth

Fig 21.5: The Opportunity to Reduce Irrigation Need and Grow Better Crops

Reclaiming Desert

Reducing and reversing desertification may not be as difficult as it might seem. Not, that is, if urban waste is available. Compost made from urban wastes will hold five to ten times its own weight of water. Dressings of 2500 to 5000 tonnes per hectare, therefore, will hold enough rainfall or irrigation if needed, to grow a crop. Oxidation of the organic matter will vary enormously, up to 100%, but it can be limited to just 10% by cropping, cultivation techniques and waste/compost technology.

The arithmetic is straight forward. In approximate terms for a semi-arid area where there is some rainfall:

- 500 tonnes of 'waste' will make around 350 tonnes of compost.
- 350 tonnes of compost per ha will hold around 2000 tonnes of water per ha (equivalent to 200 mm of rain/irrigation).
- 1 ha of crop will probably make one tonne of biodiesel and 200 litres of bioglycerol.

To reclaim hot desert, possibly double the inputs of wastes. In deciding application rates, there needs to be recognition that there is no humus to start off with and the higher temperatures will increase the oxidation rate of the organic matter. This rate will depend on ambient soil temperatures and on cultivation techniques, and these, in turn, will affect crop species choice.

What to plant? Trees give greater year-round stability to soils and the most commonly used for biofuels, so far, are palm and jatropha. However, there are others. In some areas, safflower works well and the oil has the advantage that a standard diesel engine will burn it without esterification. Indeed, diesel engine design (and more importantly, engine oil filter development) is progressing and clean oils from a variety of plant species is now possible. In some areas, where there is little or no available urban wastes, the local scrub may be composted on a rotational basis – almost any source of organic matter will do.

To make an arid soil productive, there needs to be nutrients and stability in the face of erosion – particularly wind erosion in dry times and water in a rainy season (if there is one). Wastes can be used to do that. The safety framework needed is control by agricultural scientists with local knowledge, analysis of incoming materials and the use of an Albrecht-Kinsey soil monitoring programme. With this framework arid soils can be managed to produce tree crops (including oil-producing jatropha and oil palm) and food which will contain a good range of trace elements and will probably help people live longer with less disease.

Waste has always been a political issue in every country in the world, but it is comparatively recently that *recycling* of waste has taken top of the waste agenda and the word 'recycling' has come to mean high-tech solutions and, very often, *Energy from Waste*. However, waste can be used to recover arid and desert soils, make them productive and, in so doing, remove enormous amounts of carbon dioxide out of the atmosphere and put oxygen back in.

Whole, unseparated MSW can be composted to produce a safe material which can be used as a nutrient-rich soil stabiliser[185,186,187] It may be safer and more economic to remove 'tin' cans and glass but they can be left in and the following system will still work. Cans can be reasonably easily removed by magnets which will remove the zinc used as a coating and the ferrous metal may be worth net cash (zinc is important as a trace element in the healing mechanism in human metabolism but can be toxic if too much is present). Glass may be dangerous simply because it may cause injury by cutting flesh, or focus sunlight and cause a fire. Despite these dangers, with adjusted management, it may be possible that both can be left in safely. It is also true that, in many developing areas which are also hot, then there will be a significant proportion of plastic containers in MSW. These are generally made from *polyethylene terephthalate* (PET) which may have a market value. So, the composting operation may proceed with whole MSW or with residual waste at the end of a separation process. Provided science-based judgements are made, supported by laboratory testing, there is no reason to reject this source of organic matter just because it may look unsightly. Indeed, that large particle content may be turned to advantage in stabilising the surface layer against erosion.

The nutrient value of the compost will, of course, vary with the input material. The stability of the surface mulch and how untidy it looks will, in turn, also depend on input material but also how it is treated. According to Ashish Kumar Singh[188] working with colleagues in India, MSW will contain over 75% biodegradable material and, in that, there is plenty of nitrates, potassium, magnesium and other plant nutrients.

185. Butterworth, B: *Waste in the Next Millennium, Resource, Journal of the American Society of Agricultural and Biological Engineers,* July 1998

186. Etriki, J: *Municipal Solid Waste Management and Institutions in Tripoli, Libya: Applying the Environmentally Sound Technologies (ESTs) Concept, University of Hull,* May 2013

187. Butterworth, B: *Managing the Soil Rumen and Ion Exchange, Arable Farming,* September 9th 2000

188. Kumar Singh, A et al: *Assessment of the Input of Landfill in Groundwater Supply, Environment Monitor,* #141 2008, pp309–321

In the developing countries, a large proportion of MSW is what the research describes as *biodegradable*. However, the truth is that everything is biodegradable but some materials are more easily degraded, or faster, than others. It is, in the context of this discussion, useful to distinguish between easily (such as food waste) and very lowly (such as plastic containers) degradable materials. The material which goes through the screen in the process described here can be safely used for growing food and the material going over the top of the screen can be put down the trench in the tree-growing technique. The slow-to-degrade material is exactly what is needed to stabilise the surface layer. If it has particle sizes limited by passing, say, a 100 mm screen, then it is easier to handle and mix with the topsoil and looks less untidy.

The best method of composting is to pile up in a large heap, around three metres deep and turn across a screen. What goes through the screen can probably be used as a compost for fertilising, quite safely, food crops. What goes over the top of the screen can be used for stabilising the top soil. It will look untidy but it will do the stabilising job. This can be done safely provided there is some control over what wastes are used and the output is checked by laboratory testing related to appropriate risk analysis.

Trench Reservoirs
The idea of the trench reservoir is shown in fig 21.6 below. These can be useful for establishing tress. Low grade compost can be used for stabilising buffer rows of trees on the edges of reclaiming desert.

TOP SOIL RESERVOIRS

TRENCH RESERVOIRS

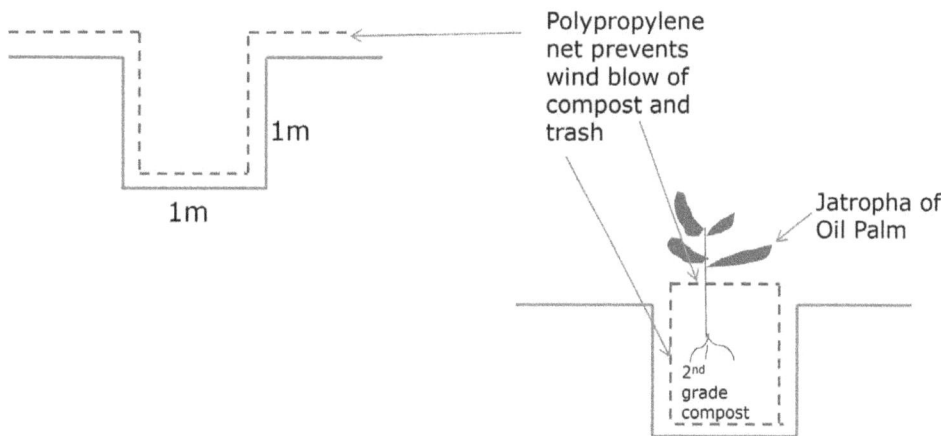

Fig 21.6:Trench Reservoirs

In extreme arid conditions, if a trench is cut, at least half a metre deep, filled with reasonably good compost and then covered with a mulch of larger particles, just mixed into the surface, say, fifty to 100 mm deep, with the trees planted though the surface layer, then that will provide a reasonable start with minimum irrigation need. What the material in the trench does is to provide a top-soil reservoir; most composts will absorb and hold five to ten times their own weight of water. What the top layer does is protect the start-up from wind erosion and limit the damage from sunlight oxidising the organic matter in the trench reservoir. An alternative top layer is one of sheet plastic (such as polyethylene – which is commonly known as *polythene*).

Lower grade composts, including those containing plastics and other large solid parties, can be used in a deeper trench as in fig 21.6 above, but using a liner of polypropylene netting which goes into the trench first, the low grade compost added and the netting closed over the top and fixed. This will stop wind-blow and allow recovery at a later date if needed. What these trenches do is provide for tree planting (such as jatropha for fuel oil for biodiesel) and windbreaks.

As this book shows, deserts can be reclaimed for crop production. Maybe not all deserts and maybe not all right now. However, the technology is there and there are real possibilities.

Why bother? Well, there is a twin pincer movement squeezing the availability of land for food production. Firstly, there is a growing population. Secondly, each day, there is less land globally. Some land goes under concrete to house the rising population. Global warming will result in loss of land to the rising sea level. So, reclaiming desert is part of the solution.

Section 6

Government & Making It happen

Chapter 22: The Part of Government

Understanding Risk

The UK *Environmental Protection Act 1990* actually says, and this is often quoted by the Environment Agency, "there must be no risk to human life, wild life or the environment". In the real world, away from the faceless people who drafted such a sentence, there is no such thing, anywhere in the universe, as 'no risk'. Doing nothing in an ever-changing environment is often a bigger risk. Running throughout this book there has been a theme of the balance of risks.

No reasonable person would ever argue that we do not need regulation in a civilised and fair society. The difficult question for a regulator is when to stop trying to restrict the bad practices, for fear of restricting the good ones. The truth is that the regulator can't win in the public eye because there will always be either pollution by bad practice, or lack of recycling because of unhelpful restrictions. Leadership can get over these difficulties by instructing that we accept some errors of certain types and accepting the flack from the media and public that follows. The British daily press has a responsibility here and that is to report in a balanced way, rather than wind up sensationalism about the negatives. There are, however, some potentially useful tools to assist those reasonable and bold enough to tackle the problem in front of us.

To be help or hindrance; that is the question.[189]

Regulations

Jim Collins, in his book[190] on his excellent research into what makes a really great business operation, made a very interesting and succinct observation about bureaucracy. The research team looked at, amongst other operations, that of George Rathmann, the co-founder of Amgen, a very successful pharmaceutical company in the USA. Collins saw that Rathmann understood that 'the purpose of bureaucracy is to compensate for incompetence and lack of discipline'. In that, the regulators clearly have a function and the state, wherever it is, has a duty to set up just such a control mechanism; there is always incompetence and indiscipline in every area of society and nowhere more than in handling wastes. However, Collins goes on to observe that what happens in most organisations is that the culture of the bureaucracy is to develop itself in order to build increasing improvements in risk management and safety levels. This is encouraged and accelerated by the small percentage of operators who are prepared to take unacceptable environmental risks and continue, in their own culture, to try to get round the regulations and the regulators. This tends to increasingly inhibit the disciplined operators who really want to 'do it right' and they get faced with a choice of bending the rules or going out of business. Rathman was spot on; the inevitable result of prescriptive regulation is to turn the good guys into bad guys.

---

189. Apologies to Shakespeare!

190. Collins, J, *Good to Great*, *Random House Business Books*, 2001

The alternative to such a negative development is that the regulator has to find a way of imposing a discipline without switching off the good guys. This is always possible if the right minds are put to work on the problem and the political will is there. There are potentially many solutions to this one is outlined below.

There is one more pitfall to be recognised before looking at the potential solution. There is a cancer-like mechanism in bureaucracy. It is a bit like coat hangers; put them in the dark and they breed. The democracies of the Western nations have a real problem and it is the growth of numbers employed in the state sector. Generally, two regulators do half as much effective policing and twice as much inhibition to good operators, as one regulator.

Most economists agree that when public expenditure (of a state) gets to 50% of total national spend (as GDP), the situation is, at best, potentially very difficult. Some economists argue that at that point, total collapse of the economy is unavoidable. The UK is close to that 50%.

The historian Jane Marshall once said, "it is in the history of the world that, whenever an empire collapses and for whatever reason, those left in power in the middle pass more and more regulations, involving more and more public servants, in order (they think) to reverse the collapse. What actually happens is that they stifle innovation and inhibit entrepreneurial activity, so accelerating the rate of decline. That is what is happening in the UK, here and now."

Alternatives to Recycle-to-Land
Politicians feel safer when passing approvals for projects which cost a lot of money, have large amounts of bright shining steel and glass, bells ringing and girls dancing. Yet there is a lamentable track record of wasting time and money. From trying to get the patient records of the NHS onto one computer, to HS2 (the high speed rail link between London and the North), there is a record of spiralling cost, delays and, too often, complete failure. Yet, so often, it appears no-one gets hung for anything. The taxpayer just coughs up. In the development of farm-scale, proximity recycling of wastes, something similar has happened. Anybody with business skills can take local costings and write a business plan for a waste-to-land recycling operation, whether it be via composting or direct spreading. If the gate fee for each material to be taken in is known, then there may be some security and value in that plan. However, with the present rapid and significant changes in regulation, technology and energy costs, long term planning is not easy unless there is a matching long term plan of supply and gate fees. This is why companies offering EfW and MBT plants, which may cost several hundred million pounds sterling, often demand twenty or thirty year contracts. The problem with such a commitment is, again, the rapidly changing business and technical environment in which we operate. It is inevitably true that these plants will become superseded and out of date before or shortly after they are commissioned, let alone before they complete their design and financial investment life. The reason for these huge EfW and MBT plants is, too often, because of political and misguided views or 'advantages of scale' and statutory targets, rather than because they offer the best long term solution both financially and environmentally. The need for flexibility in these big facilities is not easy to satisfy and it is not the main subject of this book. However, there is one real advantage of on-farm recycling to land; it is relatively flexible.

Enabling

Running an efficient and enabling civil service depends on having a very clear vision of what the objectives are and a very clear discipline of the joint obligations of policing and enabling. Generally, that cannot be achieved with more staffing, it is achieved with less staff, properly trained, properly paid and properly led. This comes down to leadership at the very top, clearly understood mission statements and backing of the people. I have a vision of the French civil service where, to get to the top, you have to be trained at La Grande Ecole, where there is one thing that is deeply entrenched in teaching which is the question; is this good for France?

Bringing this back to reversing global warming is a very difficult task for any leader.

Even slowing down global warming, never mind stopping it or reversing the process, cannot happen anywhere in the world without political decisions backed by political will on all fronts, to make it happen. Politicians are frightened that their country might lose out and they personally would be seen to be responsible. To take these risks on is a very tall order; countries are unlikely to develop a consensus for action which is active enough, until it is too late. However, there is nothing like pure, undiluted greed for driving a situation. What it comes back to is that environmental sustainability cannot be achieved without financial sustainability. If the cash is there, then the process can be driven forward against the clock. The only real hope is that there is both political will *and* entrepreneurial opportunity.

It won't happen anywhere unless it is financially sustainable. The solution posed here is financially sustainable and has been done. It is a workable, in-place solution with a practical track record. It needs scaling up but it does work.

Would it work if it were scaled up? Could it be scaled up far enough to crack a global issue of this nature? Certainly not by some super, patented gadget to be made in factories or by burying carbon dioxide in porous rock a couple of miles down. However, yes, *technically*, what is described in this book really can be done. It has been done before in the Carboniferous Era. We *know* it works.

The Structure: Can it be scaled up? Well, Chapter 17 gave a practical solution involving a mechanism to manage a landbank which can recycle wastes to land. The 'reverse franchise' mechanism developed by Land Research does actually work. Farmers in it do co-operate and they do include in that a high degree of self-regulation.

The Permit – 'Driving Licence' Permits: Land Network has suggested that a new system of permits would dramatically simplify the process of application and policing. The development would change the nature of the permit (and all the exemptions) to a format similar to the driving licence used for car drivers in the UK (which, incidentally, has the safest roads in Europe and possibly the world). Waste management facilities would get a license easily and quickly. This could be controlled by either the responsible operator passing a test or examination or having an established track record of safety and compliance with the law. Permissions could be progressive, with record earning wider permissions. Random inspections would establish if a set of simplified regulations were being adhered to and, if not, a penalty of, say, three points added to the site record. Cause pollution and the record gets, say, six or twelve points depending on significance of pollution. Get twelve points and the gate is closed to any new business for, say one week in the first instance, three months for a repeated offence.

One more idea; to provide a global mechanism to manage the emissions from aircraft, set up the United Nations to levy and collect an aviation fuel tax. There are comparatively few oil producing countries and companies and the mechanism could be manageable. It could also provide a model to build on.

The key is just one word: *enablement.*

There is nothing more difficult to carry out, nor more doubtful of success, nor more dangerous to handle, than to initiate a new order of things. For the reformer has enemies in all who profit by the old order, and only lukewarm defenders in all those who would profit by the new order. This lukewarmness arises partly from fear of their adversaries who have the law in their favour; and partly from the incredulity of mankind, who do not truly believe in anything until they have had actual experience of it.

—Machiavelli in *The Prince* (1513)

Section 7

What to Believe

Chapter 23: Discussion & Conclusions

The Triad

We, quite certainly, have a potentially very serious problem. We do not know the speed of progression and it is very complex – the most complex situation that the human race has ever event had to tackle.

There is good reason to hope, but it is mixed.

Each one of us, each organisation, each government, can, if they so wish, avoid consideration of;

   i)   How serious is global warming?

   ii) Is population growth a real danger?

   iii) What will the availability and effectiveness of vaccines and antibiotics? and

   iv) What happens if these factors, and no doubt others, become critical at the same time?

These are perfectly legitimate questions and anyone not considering them is, undoubtedly, irresponsible.

Some progress is being made:

   i)   The World Research Institute reported after the Poland conference, "tough climate talks wrap up in poland, putting Paris agreement in motion".

   ii) China did have some success in limiting birth rate. However, we, the people of the world, have got used to avoiding the issue of expanding population. Instead, we worry about how to take care of the aging population with the response of encouraging larger families.

   iii) The issue of vaccine and antibiotic failures is already causing concern that is leading to real action from the minds of the world's health ministers at governmental level.

Discussion

Can It Be Done?

We can all take notice and do something. And by the way, it will not be easy to do.

The United Nations environmental summit at Bali in December 2007 got the ball rolling, with the promise that Copenhagen in March 2009 would actually give new and real direction in control of global warming and in the reduction of greenhouse gas production. What happened in Copenhagen was that, yet again, scientists urged governments to accept that the speed of progress of global warming was greater than previously accepted by consensus. This has continued in Paris and on in Katowice, Poland ('COP24'). From these discussions, there is a potential positive which is already giving progress; there is a potentially enormous cash possibility in the form of *carbon credits*. Basically, carbon credits allow someone who can prove that they are reducing carbon dioxide in the atmosphere, to sell that evidence of carbon capture to someone, such as an industry burning fuels, who is producing carbon dioxide. The idea is to balance the two. It may be that owners of forest can offset their carbon capture capability, in a formal way, to obtain carbon credits with a substantial cash value. The question is what will be defined as 'forest'? Could it be that the definition will include palm plantations which will be fertilised with compost made from 'wastes' and planted with trees that produce liquid biofuels? There is a certain logic about this and there is only limited scope for farmers to state this case and get it onto the agenda.

Despite the hope that comes from each of these intergovernmental meetings, the *Independent* newspaper at the time reported, 'COP24 climate summit – live: UN chief warns failure to agree would be 'suicidal' after 'rogue nations' block major scientific report'. An agreement was fudged through on manipulating words in the final communique. And we move on with President Trump not being helpful. The fact is that even if the politicians do understand the threats of global warming (unlikely) they certainly do not want to be seen by the people in their own country as selling them short. Real progress is difficult at any speed and the evidence is that we do not have much time.

In the year of 2018 we learned,[191] once again, that climate change is not a distant phenomenon, it is here right now. "2018 is shaping up to be one of the hottest years on record, with new temperature records in many countries. This is no surprise. The heatwaves and extreme heat we are experiencing are consistent with what we expect as a result of climate change caused by greenhouse gas emissions. This is not a future scenario. It is happening now," said WMO Deputy Secretary-General Elena Manaenkova.

Extreme weather hit communities around the globe, ice and glaciers shrunk and global greenhouse gas emissions grew. Many of these events are in line with projections of a warming world. At the same time, our understanding of climate science greatly improved, allowing us to better understand past impacts and what the future holds.

The fact is that we really do not know at what speed global warming is occurring but there is little doubt that it really is happening and at a speed which is disturbing. Anyone who does nothing is certainly letting their children down and probably themselves. Yet the opportunity still exists to consider the options, develop new cropping and new business projects. It is the use of business that will drive this. It is only the business manager who has the skills to drive this and the opportunity is there. It is governments which can open that route and provide the framework of safety.

There is one really firm conclusion to all of the discussion about climate change;

The Only Certainty is Uncertainty
What this book has attempted to show is that urban waste can safely substitute for manufactured fertiliser and that the economics of using land as an incredibly capable recycling factory which can deliver better crop yields at lower cost and, as an added bonus, can take carbon dioxide out of the atmosphere, pump oxygen back in, and build

---

191. Levin, K & Tirpac, D: *2018: A year of Climate Extremes*, <u>World Research Institute</u>, December 27th 2018

BACS, the *Bio-Active Carbon Sink*. The principles and the technology is there. What makes this possibility a commercial success is good business management of environmental responsibility and active care.

Conclusions

There is an enormous number of books on 'the environment' and much better academic knowledge that I could ever aspire to. From childhood and later, as an agricultural scientist, I have been involved in recycling to land and, maybe thirty years ago, became involved in recycling urban and industrial wastes to farmland for food production. Early on in that development, I wanted to know what actually happened to the composted waste in the soil and if repeated additions was sustainable, i.e. could it be done for 1000 years and still do another 1000? I believe I was the first to put the jigsaw together ("I stand on the shoulders of giants.") and publish both in technical industry journals and as academic papers (see www.landresearchonline.com). What happens, is the '*closed loop*' and, if it is understood and done right, recycling of a wide range of wastes from all parts of human society, including many industrial wastes and body parts from hospitals, can be sustainable and make what may be the biggest single contribution to limiting and reversing global warming. I was responsible, despite fear based on ignorance and active restriction from the regulators, for setting up and supervising recycling over five million tonnes of many of the urban and industrial wastes listed in the *Appendix* of this book. I am of the view that the technology and practical experience that emerged must be passed on and is the basis of avoiding the tipping point trigger – because the technology links human population numbers, disease, food production, energy use, and recycling. It also covers the responsibility of individuals and of governments.

Recycling wastes to land to grow food and energy crops makes environmental and financial sense; it is very attractive. The bonus is that planting trees, preferably for energy production, on recovered land takes net, enormous amounts of carbon dioxide out of the atmosphere and pumps oxygen back in. The fact is that the whole concept is cash driven by the cash involved in handling 'wastes'. In climates or times when water is short, the obvious and emotive bonus is the dramatic reduction in irrigation need. These arguments are sound provided there are adequate, science-based controls to avoid pollution. Regulation is necessary in any society, however, the danger in regulation is over-regulation which stifles innovation and entrepreneurial activity. Looking at waste volumes globally (which are currently closely linked to wealth and therefore will grow) common sense says the only place to put most of it is on the land.

So, while I do fear, I am *not* despondent about the fact that, in my lifetime, there is certainly a threat of a disaster greater than anything seen before. In our children's lifetimes, that threat is very real but, again, I am of the view that we already have enough knowledge to progress and probably avert the worst of it, and maybe reverse it to at least a limited extent. What I am despondent about is the tendency of the human race to avoid unpleasant threats until it is too late and then try to muddle through. Thus, do I think the human race will really tackle this in an organised way and in time? Almost certainly not. However, with the right information presented with blunt truth and a credible solution? Maybe? Well, we've got to give it the best shot.

Summary

Maybe the best summary on global warming has been by Desmond Carrington, environment editor of *The Guardian* in January 2019. Carrington leads a team of probably the most balanced, professional journalists in any newspaper, on this subject. He referred to a report[192] drawn up by 54 scientists of global stature who, jointly had detailed where we all stand on global warming. Wrote Carrington:

> The level of climate-warming carbon dioxide (CO_2) in the atmosphere is forecast to rise by a near-record amount in 2019, according to the Met Office.
>
> The increase is being fuelled by the continued burning of fossil fuels and the destruction of forests, and will be particularly high in 2019 due to an expected return towards El Niño-like conditions.[193] This natural climate variation causes warm and dry conditions in the tropics, meaning the plant growth that removes CO_2 from the air is restricted.
>
> Levels of the greenhouse gas have not been as high as today for 3-5 million years, when the global temperature was 2-3°C warmer and the sea level was 10-20 metres higher. Climate action must be increased fivefold to limit warming to the 1.5°C rise above pre-industrial levels that scientists advise,[194] according to the UN. But the past four years have been the hottest on record and global emissions are rising again[195] after a brief pause.

Urgent Action

- **Stop the flippant waste of energy from hydrocarbon fuels.**
  - Get out of the car and do not buy a new one.
  - Stop looking for new experiences which involve getting on an aeroplane.
  - Turn down the central heating.
- **Stop buying stuff that is not really needed. Manufacture of anything costs energy to make and deliver.**
- **Go local including stopping commuting and taking care of our sick and old.**
- **Do not be responsible for producing waste.**
  - Do not make it in the first place.
  - Simplify manufacture so that it uses less energy and, if it produces waste, so that it can be recycled.
  - Remember that one tonne of nitrogen fertiliser nutrient, made in a modern and efficient factory, takes 21,000 kW h to manufacture and deliver – so use waste to fertilise crops and grow trees to build the global *Bio-Active Carbon Sink*.

---

192. Carrington, D: *The Guardian*, referring to *Global Carbon Budget 2018*

193. *The Guardian*

194. *The Guardian*

195. *The Guardian*

- **Exercise more to keep yourself out of hospital.**
- **Eat fewer animal products.**
- **Actively support efforts to build new technologies to help crack the problems, including:**
  - Developing new energy including tidal power and nuclear fusion.
  - Developing new vaccines and antibiotics.
- **Remember that there are many children in the world that have no parents and care for one of them rather than bringing more into the world.**
- **Be hopeful, active and happy!**

Author's Post Script

This book did not set out to crack all the problems but, in using waste at a global level, to build the global *Bio-Active Carbon Sink* (BACS), it can make a real contribution to taking carbon dioxide out of the atmosphere and giving us back the oxygen – on a large scale that would make a difference. Following on from that, are consequences for food production and human health.

Bill Butterworth
Land Research Ltd
March 2019

Final Thought

This book started with a quotation by Stephen Hawking and he shall have the last word, too.[196]

> It will take people, human beings with knowledge and understanding, to implement these solutions. Let us fight for every man to have the opportunity to live healthy, secure lives, full of opportunity and love. We are all time travellers, journeying together into the future. But let us work together to make the future a place we want to visit. Be brave, be curious, be determined, overcome the odds. It can be done.
>
> —Professor Stephen Hawking

---

196. Hawking, S: *Brief Answers to the Big Questions*, *John Murray*, London, 2018

Also from Bill Butterworth

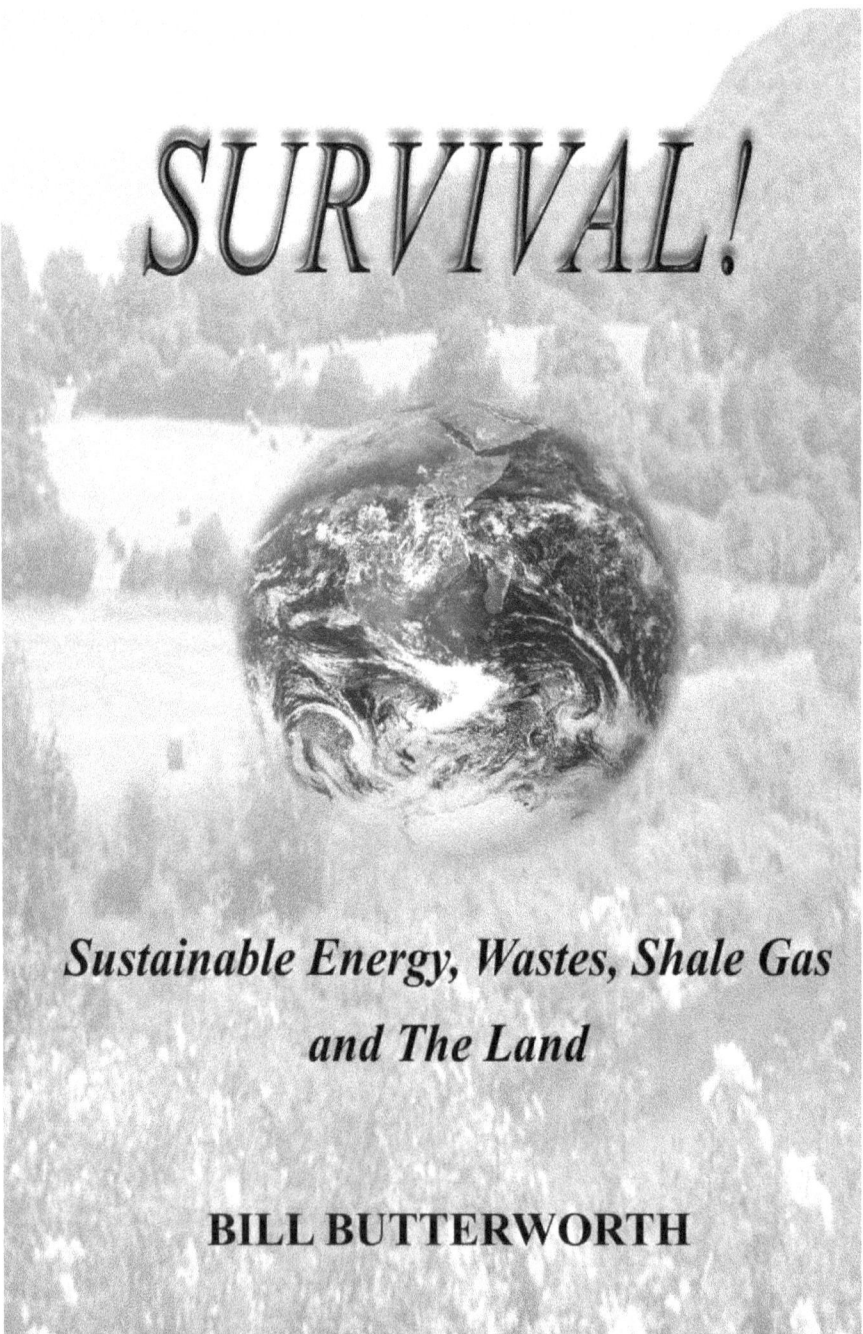

SURVIVAL!

*Sustainable Energy, Wastes, Shale Gas
and The Land*

BILL BUTTERWORTH